SpringerBriefs in Applied Sciences and Technology

Series editor

Janusz Kacprzyk, Polish Academy of Sciences, Systems Research Institute, Warsaw, Poland

SpringerBriefs present concise summaries of cutting-edge research and practical applications across a wide spectrum of fields. Featuring compact volumes of 50–125 pages, the series covers a range of content from professional to academic.

Typical publications can be:

- A timely report of state-of-the art methods
- An introduction to or a manual for the application of mathematical or computer techniques
- A bridge between new research results, as published in journal articles
- A snapshot of a hot or emerging topic
- An in-depth case study
- A presentation of core concepts that students must understand in order to make independent contributions

SpringerBriefs are characterized by fast, global electronic dissemination, standard publishing contracts, standardized manuscript preparation and formatting guidelines, and expedited production schedules.

On the one hand, **SpringerBriefs in Applied Sciences and Technology** are devoted to the publication of fundamentals and applications within the different classical engineering disciplines as well as in interdisciplinary fields that recently emerged between these areas. On the other hand, as the boundary separating fundamental research and applied technology is more and more dissolving, this series is particularly open to trans-disciplinary topics between fundamental science and engineering.

Indexed by EI-Compendex and Springerlink.

More information about this series at http://www.springer.com/series/8884

Iraj Sadegh Amiri · Mahdiar Ghadiry

Analytical Modelling of Breakdown Effect in Graphene Nanoribbon Field Effect Transistor

 Springer

Iraj Sadegh Amiri
Computational Optics Research Group
Ton Duc Thang University
Ho Chi Minh City
Vietnam

Mahdiar Ghadiry
Photonics Research Centre
University of Malaya
Kuala Lumpur
Malaysia

and

Faculty of Applied Sciences
Ton Duc Thang University
Ho Chi Minh City
Vietnam

ISSN 2191-530X   ISSN 2191-5318   (electronic)
SpringerBriefs in Applied Sciences and Technology
ISBN 978-981-10-6549-1   ISBN 978-981-10-6550-7 (eBook)
https://doi.org/10.1007/978-981-10-6550-7

Library of Congress Control Number: 2017955270

Printed on acid-free paper

This Springer imprint is published by Springer Nature
The registered company is Springer Nature Singapore Pte Ltd.
The registered company address is: 152 Beach Road, #21-01/04 Gateway East, Singapore 189721, Singapore

Contents

About the Authors

Dr. Iraj Sadegh Amiri received his B.Sc. in applied physics from the Public University of Urmia, Iran, in 2001 and a gold medal in M.Sc. from the University Technology Malaysia (UTM) in 2009. He was awarded a Ph.D. in photonics in January 2014. He has published well over 350 academic journal/conference papers and books/chapters on optical soliton communications, telecommunications, fibre lasers, laser physics, waveguide fabrication and application in photonics, photonics, optics, nonlinear fibre optics, quantum cryptography and bioengineering. He was a Junior Researcher at the University Technology Malaysia (UTM), Laser and Photonics Center, and a postdoctoral research fellow at the University of Malaya (UM), Photonics Research Center (PRC). Currently, he is a Senior Lecturer at the University of Malaya (UM), Photonics Research Center (PRC), under the directorship of Prof. Dr. Harith Ahmad.

Mahdiar Ghadiry is a postdoctorate in electronics at the University of Malaya (UM), received his Ph.D. in microelectronics and has more than 6 years of experience at the university including managing, lecturing and supervising master's and degree-level students. He has an extensive research background and has published more than 30 ISI journal articles and three books. In addition, he has been involved in the electronics industry for 3 years and has more than 4 years of experience in integrated circuit (IC) design and embedded system design as an employee of the Electronic Components Industries (ECI), which is the biggest IC design company in Iran.

Abstract

Since 2004, graphene as transistor channel has drawn huge amount of attention due to its extraordinary scalability and high carrier mobility. In order to open required bandgap, its nanoribbon form is used in transistors. Breakdown effect modelling of the graphene nanoribbon field-effect transistors (GNRFETs) is needed to investigate the limits on operating voltage of the transistor. However, until now there is no study in analytical approach and modelling of the breakdown voltage (BV) effects on the graphene-based transistors. Thus, in this project, semianalytical models for lateral electric field, length of velocity saturation region (LVSR), ionization coefficient (α) and breakdown voltage (BV) of single- and double-gate graphene nanoribbon field-effect transistors (GNRFETs) are proposed. As the methodology, the application of Gauss's law at drain and source regions is employed in order to derive surface potential and lateral electric field equations. Then, LVSR is calculated as a solution of surface potential at saturation condition. The ionization coefficient is modelled and calculated by deriving equations for probability of collisions in ballistic and drift modes based on lucky drift theory of ionization. Then the threshold energy of ionization is computed using simulation, and an empirical equation is derived semianalytically. Finally, avalanche breakdown condition is employed to calculate the lateral BV. As a result of this research, simple analytical and semianalytical models are proposed for the LVSR, and BV, which could be used in design and optimization of semiconductor devices and sensors. The proposed equations are used to examine the BV at different situations of various channel lengths, supply voltages, oxide thickness, GNR's widths and gate voltages. Simulation results show the operating voltage of FETs could be as low as 0.25 V in order to prevent breakdown. However, after optimization, it can be reached to 1.5 V.

Chapter 1
Introduction on Scaling Issues of Conventional Semiconductors

Abstract In this section, firstly, a brief background is presented to explain the issues connected with CMOS scaling and breakdown voltage. Secondly, the research objectives, scope, plan and a brief methodology of this project are expressed.

Keyword Scaling issues · High voltage problems · Graphene

1.1 Background of CMOS Scaling Problems

Metal oxide field-effect transistor (MOSFET) as shown in Fig. 1.1 has been the most used semiconducting device for low-power logic circuits, power MOSFETs and analogue applications. The key advantages of MOSFET compared to previous counterparts such as resistor–transistor logic (RTL) and bipolar junction transistor (BJT) are its low power consumption and high input impedance due to isolation of gate from channel. However, high delay of CMOS (complementary MOS) used in digital applications has been always an issue compared to high switching frequency of, for example, BJT logics.

For decades, there has been a lot of improvements in lowering power and delay in MOSFETs by changing the gate dielectric, altering the structure and using different layers, adding several gates leading to double gate, triple gate and even surrounding gate MOSFETs to control the channel better and obviously employing different channel material such as GaAs instead of silicon to increase the carrier velocity.

Alternatively, shrinking transistor sizes has been one of the most significant solutions for improving power-delay product (PDP). Reducing the channel length, results in lowering the channel resistance and delay. In addition, it causes the gate capacitance, which is the most important factor in logic gates' delay, to reduce *International roadmap for semiconductor technology* (*ITRS*). There are limitations such as short channel effects which prevent scaling down to nanoscale dimensions and reaching desired characteristics.

© The Author(s) 2018
I.S. Amiri and M. Ghadiry, *Analytical Modelling of Breakdown Effect in Graphene Nanoribbon Field Effect Transistor*, SpringerBriefs in Applied Sciences and Technology, https://doi.org/10.1007/978-981-10-6550-7_1

Fig. 1.1 Conventional
MOSFET with isolated gate
from channel using oxide.
Each FET consists of four
main parts, drain, source, gate
and channel. Gate is
responsible to control
conductivity of the channel
and establish current flow
between drain and source

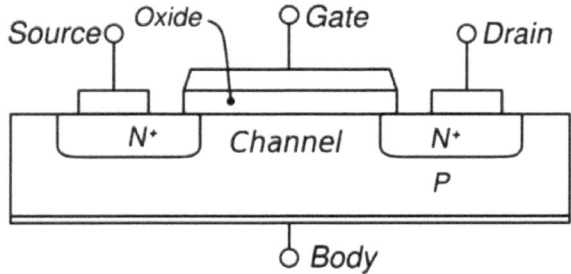

When the channel length is comparable to the depletion region of the source and drain, the device is called a short-channel device. In short channel devices, short channel effect arises that limits the device performance. Therefore, researchers have been trying to introduce new materials with higher mobility and scalability. In 2004, Geim and Nikolove [1] managed to produce stable graphene—one atom thick layer of graphite at room temperature—and measure its mobility. As it was expected from previous theoretical studies, high carrier mobility was measured in graphene, which is a promise for future nanoelectronic devices. In addition to very high carrier velocity, it shows very high conductance and tunable bandgap. However, the main issue with graphene is its zero bandgap which makes it a very poor semiconducting material for application of FETs. Further studies on opening bandgap in graphene, resulted in introduction of Carbon Nanotube (CNT) and graphene nanoribbon (GNR). Figure 1.2 shows typical samples of GNR and CNT.

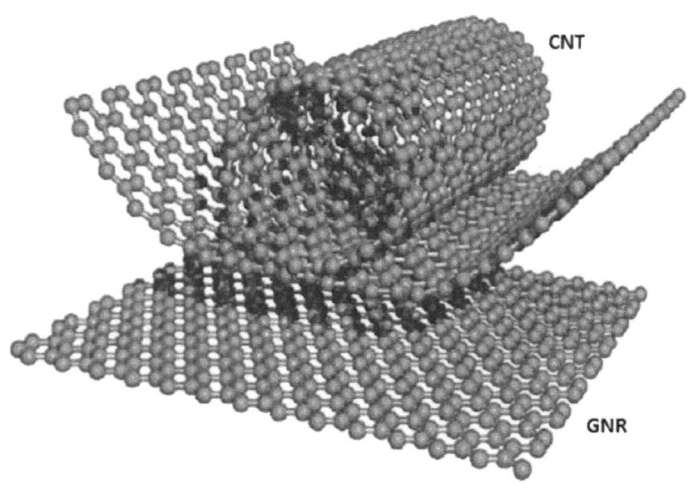

Fig. 1.2 Graphene in form of a tube is called carbon nanotube (CNT). Narrow sheet of graphene which is unzipped CNT is known as graphene nanoribbon (GNR)

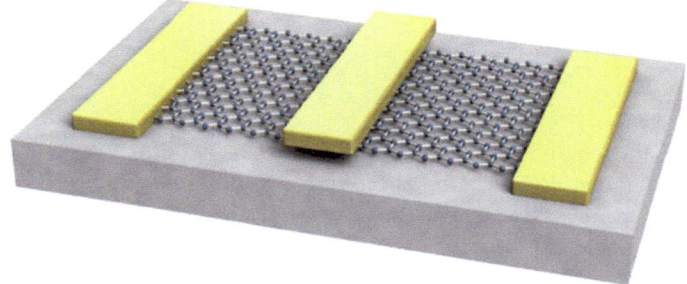

Fig. 1.3 Typical GNRFET with top gate and Au drain and source contacts. Graphene nanoribbon is used in channel to decrease the transistor switching time

Graphene nanoribbons are strips of graphene with narrow width normally less than 50 nm indicating notable electrical properties such as high mobility, high conductance and small bandgap [1]. Recently, GNR has been introduced as an alternative material for the next generation of MOSFETs [2]. Figure 1.3 shows a typical graphene nanoribbon FET (GNRFET) with a top gate. Using graphene with thickness as low as possible, the adverse short channel effects in silicon-based MOSFETs could be solved. Therefore, the dimensions of the transistors could be scaled down extremely, which results in low propagation delay down to 0.025 ps [3].

However, the benefits of GNR come with cost. Firstly, the bandgap opened in GNR is still not enough to secure a satisfactory I_{on}/I_{off}, and in narrow ribbons, edge effects suppress the mobility to some values even less than that of silicon counterpart. Secondly, fabrication of GNR is still a difficult and not accurate task [2]. Despite great improvement in fabrication process of GNR, it is still not mature enough to be used in mass production and industry. However, the research is still vastly going on in this field hoping to find solutions for these issues.

Due to difficulties in fabrication of GNR, many researchers take advantage of analytical modelling and computer simulation to extract details about properties of GNR and possibility of making applicable FETs using GNR. As a result, there are several models for properties of GNR and CNT in the literature. However, since graphene as channel material was introduced recently, there are still many unanswered questions to be explored on these materials. As an example, there has been no attempt to study the breakdown mechanism and ionization process of GNR analytically or experimentally.

Lateral breakdown, which will be the focus of this thesis, is a mechanism limiting the maximum voltage that can be tolerated before the beginning of large current flow between the drain and source in a FET. Prior to calculation of the lateral breakdown voltage, impact ionization rate must be computed. Equation (1.1) shows the relation of impact ionization and breakdown voltage [4].

$$1 = \int_0^{L_\mathrm{d}} \alpha \mathrm{d}x, \qquad (1.1)$$

where L_d is the length of saturation velocity region—a portion of channel between pinch-off point and drain—and α is the impact ionization which is the number of electron–hole pairs created by a mobile carrier travelling a unit of distance along the lateral electric field [5].

When a sufficient electric field is applied between drain and source, mobile carriers gain enough energy to create electron–hole pairs by colliding with lattice atoms resulting in impact ionization [6, 7]. This process (impact ionization) defines the current which flows in the depletion region when a large electric field is applied.

In this thesis, a study on effects of lateral breakdown voltage of GNR-based FETs is conducted. As a result of this thesis, several analytical models are proposed for breakdown mechanism and safe operating voltage of typical devices is calculated analytically. In addition, future studies on design and optimization of related devices such as power FETs or avalanche photodiodes (APDs) could use the proposed approach here.

1.2 High-Voltage Scaling Issues

Increasing the drain-source voltage (V_ds) in FETs causes the drain-source current (I_ds) to increase. However, there is a limit (breakdown voltage (BV)) in increasing V_ds. After that limit, the device does not function properly and either it conducts high amount of current or cut the current both being a failure in a circuit. Therefore, it is necessary to identify BV of any new material in the devices in order to limit the operating voltage. While in carbon-based FETs, which is the most important device in carbon-based digital and analogue circuits, there is shortage of research on breakdown voltage. Therefore, it was a motivation for us to examine the breakdown and ionization mechanisms in GNRFETs. In this project, an analytical approach is presented to calculate maximum operating voltage of GNRFETs.

1.3 Study Limitations in This Book

As fabrication of carbon-based devices requires sophisticated equipment such as advanced and accurate CVD (Chemical vapour deposition) machine and precise photo-lithography, fabrication is not possible with the available equipment in our university. Therefore, our research is limited to analytical models and computer simulations only. We only address lateral breakdown and ionization. In addition, among variety of devices such as bilayer-GNRFET and CNT-FET, we limit this project to mono-layer GNRFET for simplicity to make sure that we can achieve our objectives. However, both single-gate and double-gate FETs are modelled, and breakdown voltage is calculated.

1.4 Study Objectives on Graphene Field-Effect Transistors

- To propose analytical models for lateral electric field and length of velocity saturation region of GNR-based FETs
- To propose an analytical model for ionization coefficient and breakdown voltage of GNR-based FETs
- To simulate GNR-based FETs in terms of breakdown voltage and calculate the maximum operating voltage of the typical GNRFETs at different conditions.

1.5 Summary of Methodology Used to Study Breakdown in Graphene-Based Transistors

The modelling in this project is divided into three different sections. The first section deals with surface potential, lateral electric field and length of velocity saturation region. The second section provides models for ionization coefficient, and in the last section, the model for breakdown voltage is provided.

1.5.1 Length of Saturation Velocity Region

Surface potential will be modelled using application of Gauss's law at drain and source regions of graphene nanoribbon channel. As Fig. 1.4 shows, the models are derived using one-dimensional approach for simplicity. Firstly, we start by applying Gauss's law inside the channel to obtain Poisson's equation. Then surface potential is resulted by solving the Poisson's equation. By taking derivation, lateral electric field can be obtained. In addition, using the surface potential expression, the length of velocity saturation region is achieved.

1.5.2 Impact Ionization Coefficient

Impact ionization model can be derived based on general lucky drift theory reported in [8] and successfully used for semiconductors with parabolic bandstructure such as Si and GaAs [5]. In this method, it is assumed that a carrier can reach threshold

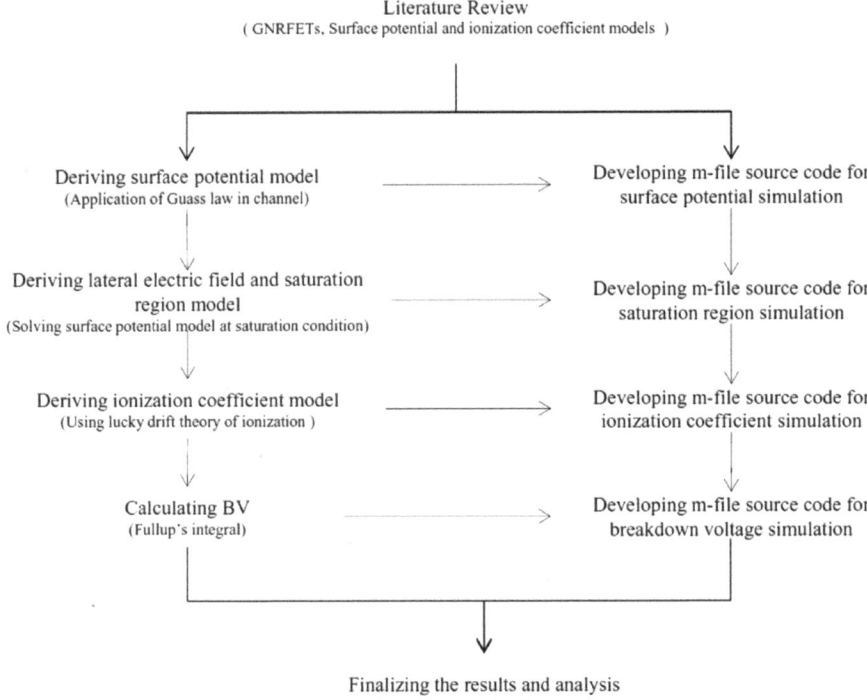

Fig. 1.4 Flow chart is used to conduct this book

energy in two ways. First, it reaches threshold energy through a ballistic motion. Second, the carrier first undergoes some collisions and then reaches the threshold energy. Therefore, the motion of electron is modelled in both drift and ballistic modes. First, an expression for characteristic length being the distance carriers travel before reaching threshold energy having no collision is derived. Then the probability of having no collision travelling characteristic length in both ballistic and drift modes is formulated. Adding two probabilities gives the total probability of reaching threshold energy. It is worth to mention that due to unusual properties of GNRs, significant modification must be made to the previous models, which are discussed in the relevant section.

1.5.3 Breakdown Mechanism in Field-Effect Transistors

Finally, the breakdown voltage is modelled. The model relies on Fullop's integral, which has been used many times for calculation of BV in silicon-based transistors

[4]. In this method, firstly multiplication factor is calculated and then by equating the multiplication factor to infinity (avalanche breakdown condition), BV is calculated. The drain-source voltage is increased until the avalanche condition is satisfied. The obtained V_{ds} is called breakdown voltage resulting in infinite multiplication factor. In summary, a flow chart shown in Fig. 1.4 is used to conduct this project.

1.6 Book Organization

This thesis is organized as follows. Chapter 2 provides the basic concepts regarding the length of saturation velocity region, ionization mechanism and lateral breakdown voltage. Furthermore useful equations and definitions will be provided there. In addition, more information will be given focusing on the advantages and disadvantages of graphene, application of graphene in FETs and required equations and properties used in this thesis. Chapter 3 will review literature in three sections, surface potential models, ionization coefficient models and graphene-based transistors. The methodology is presented in three sections of Chap. 4 consisting three types of analytical models.The next chapter presents the simulation results based on the proposed models at different values of structural parameters. A comparison between double gate (DG) and single gate (SG) will be conducted as well. Chapter 5 presents a summary of this thesis and outlines the achieved results and recommends possible future works.

References

1. K.S. Novoselov, A.K. Geim, S. Morozov, D. Jiang, Y. Zhang, S.A. Dubonos, I. Grigorieva, A. Firsov, Electric field effect in atomically thin carbon films. Science **306**(5696), 666–669 (2004)
2. F. Schwierz, Graphene transistors. Nat. Nanotechnol. **5**(7), 487–496 (2010)
3. R. Sako, H. Hosokawa, H. Tsuchiya, Computational study of edge configuration and quantum confinement effects on graphene nanoribbon transport. Electron Device Lett. IEEE **32**(1), 6–8 (2011)
4. W. Yang, X. Cheng, Y. Yu, Z. Song, D. Shen, A novel analytical model for the breakdown voltage of thin-film SOI power MOSFETs. Solid-State Electron. **49**(1), 43–48 (2005)
5. O. Rubel, A. Potvin, D. Laughton, Generalized lucky-drift model for impact ionization in semiconductors with disorder. J. Phys. Condens. Matter **23**(5), 055802 (2011)
6. H. Wong, Drain breakdown in submicron MOSFETs: a review. Microelectron. Reliab. **40**(1), 3–15 (2000)
7. I.-J. Kim, S. Matsumoto, T. Sakai, T. Yachi, Analytical approach to breakdown voltages in thin-film SOI power MOSFETs. Solid-State Electron. **39**(1), 95–100 (1996)
8. W. Fawcett, A. Boardman, S. Swain, Monte Carlo determination of electron transport properties in gallium arsenide. J. Phys. Chem. Solids **31**(9), 1963–1990 (1970)

Chapter 2
Basic Concept of Field-Effect Transistors

Abstract In this chapter, first the basic concept of FETs is introduced. In addition, in three subsections, the concepts related to the length of saturation velocity region, impact ionization and lateral breakdown are discussed. Finally, graphene is introduced as a candidate for transistor channel and its properties related to FET are studied.

Keywords FET · Ionization · Length of velocity saturation region · Carbon-based devices

2.1 Field-Effect Transistors (FETs) and Its Issues

A FET, shown in (Fig. 2.1), is simply a device consisting of a gate, a channel region which connects the source and drain junctions, and a barrier which separates the channel from the gate. By controlling the channel conductivity in FETs the drain current increases or decreases. The channel conductivity varies by changing the applied voltage between gate and source. A threshold voltage V_t is defined in FETs as the minimum voltage of gate-source to form a conducting channel between drain and source.

There are three main regions in each voltage transfer characteristic, cut-off, linear and saturation. In cut-off state, where $V_{gs} < V_{th}$ no conducting channel is formed and therefore no current flows. In the linear region, $V_{gs} > V_{th}$ and $V_{ds} < V_{sat}$, where V_{sat} is the drain saturation voltage. In this region as V_{gs} increases, the current too increases, almost linearly respect to V_{gs}. The last is saturation region (see Fig. 2.2), where as V_{ds} increases current increases slightly.

In this region, carriers' speed reaches velocity saturation υ_{sat} and does not exceed that due to collisions, which deviate carriers from lateral direction and reduces their velocity.

© The Author(s) 2018
I.S. Amiri and M. Ghadiry, *Analytical Modelling of Breakdown Effect in Graphene Nanoribbon Field Effect Transistor*, SpringerBriefs in Applied Sciences and Technology, https://doi.org/10.1007/978-981-10-6550-7_2

Fig. 2.1 Conventional FETs. Schematic cross section of an n-type bulk silicon FET (extracted from [1])

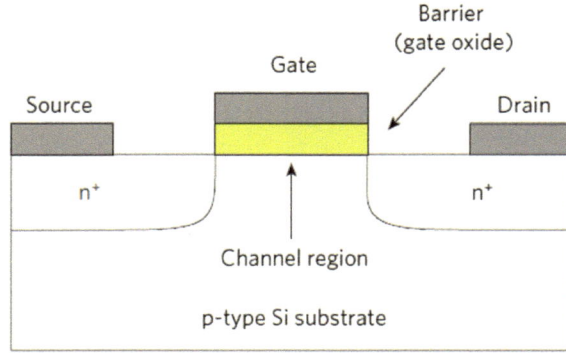

Fig. 2.2 FET transfers characteristics showing I_D against the gate-source voltage, V_{gs}. Increasing V_{ds} causes the current to increase. However, after a certain V_{ds}, which is called saturation voltage (V_{th}) a saturation point is reached and the current does not increase as V_{ds} increases

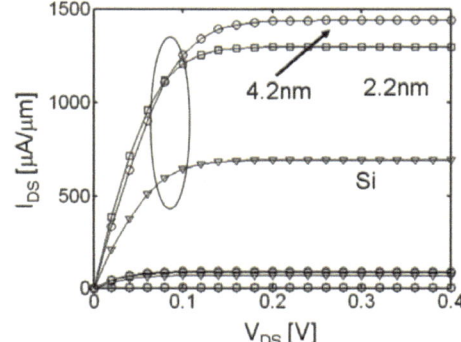

2.2 Length of Velocity Saturation Region

The effective channel length is one of the most important parameters of MOSFETs showing the portion of the channel that contribute to the properties of the MOS such as current–voltage (I–V) characteristic. In order to calculate effective channel length, which is $L_E = L - L_d$, the length of the drain region L_d has to be computed. The L_d controls the lateral drain breakdown voltage [2], substrate current, hot electron generation [3] and drain current at the drain region [4]. In a FET, if the applied drain voltage is higher than the drain saturation voltage, the electric field near the drain junction will be higher than the critical field strength, which results in carrier velocity saturation. In addition, high electric field near the drain junction causes impact ionization [5]. Saturation region is defined as the region between pinch-off point and drain (see Fig. 2.3).

As reported in [2, 6], the length of this region is used along with Fulop's Integral to calculate breakdown voltage (BV) in FETs. In high power devices, a drift region is normally formed outside the gate area to increase the breakdown voltage and length of saturation region is approximated to the length of drift region [7, 8]. Figure 2.4 shows a schematic view of a typical power device. In this figure, the length of velocity saturation region L_d and the effective channel L_E separated by pinch-off point are shown.

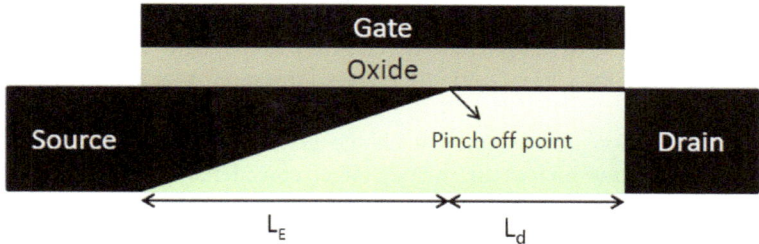

Fig. 2.3 Length of velocity saturation region L_d and pinch-off point. At high electric field, carrier's velocity reaches a saturation velocity and current saturates. Impact ionization occurs in the region between pinch-off and drain

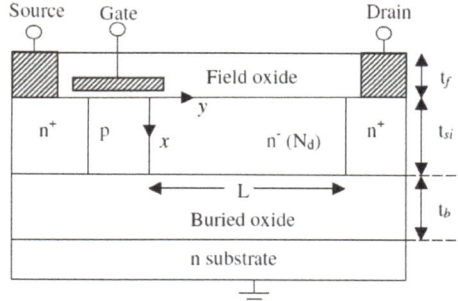

Fig. 2.4 A typical power transistor with drift region outside gate area. The t_f, t_b, t_{si} are front oxide, back oxide, channel thickness, respectively, and L is the length of drift region or L_d. In conventional power devices, increasing drift region length (L) causes the breakdown voltage to increase (figure has been extracted from [9])

2.3 Impact Ionization

As the feature size of integrated MOS devices decreases further, the high electric field near the drain region becomes more crucial and poses a limit on the device operations, notably by a large gate current, substrate current and substantial threshold voltage shift, hot-electron generation and drain breakdown caused by the impact ionization in the high-field region near the drain. The key parameters for describing these mechanisms are the impact ionization rate and the length of velocity saturation region.

The definition of impact ionization is the number of electron–hole pairs created by a mobile carrier travelling unit of distance through the depletion region along the direction of the electric field [10]. According to several previous works such as [10], the electrons and holes impact ionization coefficients are strongly dependant on the electric field strength. It can be formulated as the inverse of the average distance travelled by a carrier prior to the ionization event, and it is given by $\alpha = P(F, E_t)/l_0$, where $P(F, E_t)$ is the probability that electron reaches threshold energy E_t defined as minimum energy required to free an electron [11]. In this equation, α is the impact ionization coefficient of GNR, F is the electric field strength and $l_0 = E_t/qF$ is the distance travelled by carrier prior to impact ionization assuming no collision is possible.

Impact ionization is an important charge generation mechanism. It occurs in many semiconductor/devices and it may either considered as beneficial characteristic of the device or it can result in unwanted parasitic effects [12]. For example, it is exploited in avalanche photodiodes (APDs). An avalanche photodiode (APD) is light-sensitive electron device employing the photoelectric effect to interpret the intensity of the light to electricity. Applying high reverse bias (typically 100–200 V in silicon) results in a gain (roughly 100) caused by impact ionization and avalanche phenomenon.

2.4 Lateral Breakdown in Field-Effect Transistors

One of the most important and unique properties of power devices is their capability to resist high voltages and currents [2, 8]. In the design of transistors used for digital applications, reducing power consumption and increasing the performance are the two important objectives. One of the most influential parameter in reducing power is lowering the supply voltage [13]. In contrast, in power devices, such as transistors used to derive electric motors, the operating voltage is much higher than that of digital applications. Therefore, high breakdown voltage is required. Based on the application, the BV could be varied from around 20 to 30 V for voltage regulators used in power supply circuits in order to supply voltage for processors to over 5000 V for devices, which is employed in power transmission lines [2]. However, in nanotransistors, this voltage decreases down to even less than 2 V [13].

Tolerating high voltages without showing high and uncontrolled current flow in a semiconducting device is ruled by the avalanche breakdown related to the lateral electric field in the device [14]. Normally high electric field is seen inside the structure of the device or at the edges [15]. Therefore, the device is optimized to tolerate high drain-source voltages while the on-state voltage drop must be kept as low as possible in order to reduce the power dissipation [15].

2.4.1 Multiplication Coefficient and Ionization Integral

The condition for occurring avalanche breakdown is met if the rate of the impact ionization becomes infinite. If the electric field is increased enough, it reaches a certain level, where the carriers could be accelerated and finally gain enough energy to generate electron–hole pairs by colliding to lattice atoms. According to definition of the impact ionization coefficient, any hole creates [α_p dx] pairs of electron–hole by travelling dx in the depletion region. Concurrently, the electron does the same and creates $[\alpha_n \, \mathrm{d}x]$ pairs travelling the distance $\mathrm{d}x$. Therefore, $M(x)$, which is known as the multiplication coefficient, defined as the number of electron–hole pairs generated by a single electron–hole pairs firstly created at a distance x from the source junction, is written by Baliga [16] as

$$M(x) = 1 + \int_0^x \alpha_n M(x)\mathrm{d}x + \int_x^{L_d} \alpha_p M(x)\mathrm{d}x \tag{2.1}$$

where can be written by Baliga [16] as

$$M(x) = M(0)\exp\left(\int_0^x (\alpha_n - \alpha_p)\mathrm{d}x\right) \tag{2.2}$$

where $M(0)$ is the total number of electron–hole pairs at the edge of the depletion region, and α_n and α_p are ionization coefficients of electrons and holes, respectively. Applying this equation in 2.1 and taking $x = 0$ gives a solution of $M(0)$ [16].

$$M(0) = \left(1 - \int_0^{L_d} \alpha_p \exp\left(\int_0^x (\alpha_n - \alpha_p)\mathrm{d}x\right)\mathrm{d}x\right)^{-1} \tag{2.3}$$

$$M(x) = \frac{\exp\left(\int_0^x (\alpha_n - \alpha_p)\mathrm{d}x\right)}{1 - \int_0^{L_d} \alpha_p \exp\left(\int_0^x (\alpha_n - \alpha_p)\mathrm{d}x\right)\mathrm{d}x} \tag{2.4}$$

This equation is useful for calculation of the total number of electron–hole pairs caused by the creation of a single electron–hole pair at a distance x from the junction provided that the lateral electric field strength and distribution (in transistors) is calculated. The avalanche breakdown condition, which is met when the total number of generated electron–hole pairs in the depletion region is almost infinite, can be interpreted as the M almost equal to infinity. This condition is met by assuming the dominator of Eq. 2.4 to 0.

$$\int_0^{L_d} \alpha_p \exp\left(\int_0^x (\alpha_n - \alpha_p)\mathrm{d}x\right)\mathrm{d}x = 1 \tag{2.5}$$

The left-hand side expression is referred as ionization integral. In the calculation of breakdown voltage and analysis of the power devices, it is common to find a voltage at which make the ionization integral equal to 1 [9]. Considering equal coefficient for impact ionization of holes and electrons, the avalanche breakdown condition can be written as [2, 17]

$$\int_0^{L_d} \alpha dx = 1 \qquad (2.6)$$

Using this equation, in order to find avalanche condition and breakdown voltage, we need to calculate ionization coefficient α and L_d. This matter will be addressed using semi-analytical approaches in the following chapters.

2.4.2 Avalanche Breakdown

Electrons and holes that enter the depletion layer are swept out by the electric field within the depletion region, leading to acceleration of the carriers to high velocities until they reach saturation velocity. If the channel is made of silicon, the saturation drift velocity is about 1×10^7 m/s, which is attained at the electric field more than 1×10^5 cm^{-1} [2]. If the electric field increases even more, the mobile carriers can obtain enough energy so that their collision with lattice atoms could free an electron from the valence band and elevate that to the conduction band resulting in generation of an electron–hole pair [18]. Then the created electrons and holes, which are experiencing the electric field, contribute in further impact ionization and produce even more pairs. As a result, it is said that impact ionization is a self-progressive (multiplicative) phenomenon, leading excessive mobile carriers, which participate in flowing significant current between drain and source. As the MOSFET is not able to resist the applying higher voltages, due to a rapid increase in the current, the breakdown voltage is known as a limit for operating voltage of MOSFETs [8].

Figure 2.5 shows breakdown mechanism due to impact ionization process.

Fig. 2.5 Avalanche breakdown and substrate current in a typical FET. Impact ionization results in substrate current, which is undesired characteristic in conventional FETs (extracted from [2])

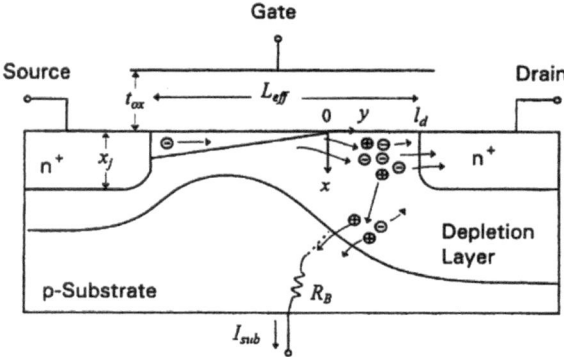

2.5   Down Scaling Problems

The performance and power consumption of digital logic rely on almost completely on the efficiency of a single device, which is the MOSFET. As mentioned before, for decades, scaling down the MOSFETs has been the most important action to succeed in digital logic. This miniaturization has made it possible that the complexity of integrated circuits (ICs) doubles each 18 months as shown in Fig. 2.6, resulting to essential progress in speed and decreases in power consumption and price per transistor. Nowadays, processors employing two billion FETs, many of them using gate lengths only 30 nm or less, are being produced (Fig. 2.6).

Moore's law has forecast the trend of silicon chips in the last forty years [19]. For more than four decades, silicon has been the most important CMOS technology of the today's information society. It is thought that silicon is going to be the dominant process for at least one more decade [19]. However, as transistor dimensions approach few nanometres the silicon transistors' behaviour becomes more uncertain making silicon improper technology for the future circuit's unless new solutions are found to address its issues [20].

For decades shrinking the dimensions of the channel, oxide thickness and operating voltage, has been the most important key to improve the power consumption and performance of the FET devices, especially in logic applications. However, this scaling cannot be continued forever as it has been anticipated several times. After years of threshold voltage downscaling, leakage current has increased from $<10^{-10}$ amp/mm to $>10^{-7}$ amps/μm. Thus, it is difficult to further lower the threshold voltage and therefore, the operating voltage cannot be reduced as well [21].

Another issue arises from scaling the oxide thickness. Although reducing the oxide thickness results in device performance improvement and operating voltage decrease, due to leakage current, it reaches the limits. Gate oxide in 65 nm technology of Intel FETs (SiO_2) is only 1.2 nm, which is equal to five layers of silicon atoms. This shows that downscaling is reaching the dimension of atoms. In other

Fig. 2.6 Trends in the number of transistors per digital chips and transistor channel. To keep up with this trends length of channel in transistors has been reduced. However, this shrinking cannot continue for too long, which is why new structures such double-gate FETs, and new materials like graphene have been introduced hoping to reach even shorter length and higher processing speed (extracted from [1])

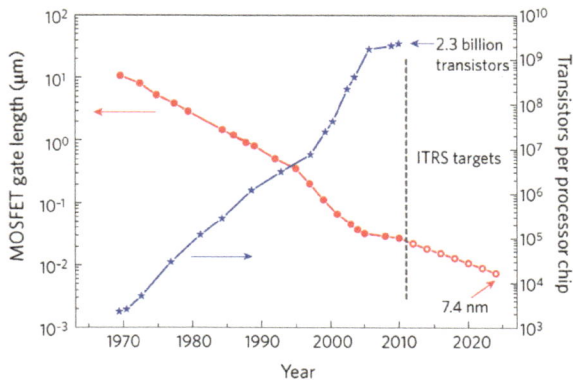

word, we are running out of atoms. Furthermore, there is a limit for increasing the doping concentration. As the doping concentration increases, the carrier velocity degrades due to increase in scattering. Reducing channel length has been also another key approach to improve characteristics of FET devices. In high-performance applications, FETs must quickly respond to V_{gs} variations, requiring high mobility and short channel. However, short-channel length results in problems such as threshold voltage roll-off and drain-induced barrier lowering (DIBL) [19].

Short-channel problems (effects) are one of the most challenging issues in the nanoscale MOSFETs. When the channel length is comparable to the junction thickness, which is relevant in nanotransistors, the gate barrier height is lowered, which leads to decreasing the threshold voltage (V_{th}). In addition, if high voltages for drain junction are applied to a short-channel transistor, the gate barrier height decreases even more, which causes the threshold voltage to decreases further. This issue is known as drain-induced barrier lowering (DIBL). Eventually, the MOSFET reaches a point called the punch-through, where the gate is totally unable to control the drain-source current flow.

Normally, two physical phenomenon are attributed to the short-channel effects, which are (1) impairing the drift characteristics of the electron in the short channel. (2) The threshold voltage changes because of channel length shortening. In other point of view, short-channel effects are distinguished into five different effects

1. Hot electrons
2. Velocity saturation
3. Surface scattering
4. Impact ionization
5. DIBL and punch-through.

According to prediction of scaling theory [21], in order to make a robust FET against short channel effects, a FET with a thin gate-controlled region (measured in the vertical direction) and a thin barrier must be designed. The fact that in graphene, it is possible to have channels that are as thin as one atom layer is perhaps the most interesting properties of graphene for application in transistors [1].

Although there are reported devices with extremely thin channels, such as iii–v HEMTs with typical channel length of 10–15 nm and silicon-on-insulator MOSFETs using channel with thickness of less than 2 nm, the rough surface results in deteriorated mobility [22]. More importantly, a significant threshold voltage variation is seen in these devices because there is a fluctuation in body thickness of these devices, and the same problem is expected to happen when the thickness of iii-v HEMT is reduced to only a few nanometres [1]. These issues are seen at thicknesses that are much greater than that of graphene.

Another important issue in the modern MOSFETs is the series resistance between the source and drain junctions, which is becoming more significant as the gate length is reduced [19]. Therefore, significant amount of research has been devoted to suppressing the short-channel effects and optimizing the series resistance in modern transistors. As a result device engineers have been trying to find alternatives materials with better scalability and higher carrier velocity [1]. So far graphene has been shown to have very high carrier velocity and scalability compared to silicon and other counterparts such as GaAs.

2.6 Carbon-Based Semiconductor Devices

As the end of silicon scaling has been predicted number of times due to technical reasons and scaling alone only results in fulfilling the needs of one generation, introducing a fundamentally new material based on essentially different physical properties compared to the silicon is of a great interest among the device engineers. However, switching to a new material is a challenging task to do. Because logic circuit fabrication needs complex processes and device fabrication plants are extremely expensive to implement. In addition, introducing new material requires the fabrication plants to be replaced or modified significantly, which costs a lot of money.

Therefore, there are objections among logic designers against introducing alternatives for silicon. However, the conditions are not the same for radiofrequency applications. This field is supported and dominated by defence applications. Because of need and advances in wireless communications, the military is wiling to spend great amount of money in research into new radiofrequency devices. In addition, radiofrequency chips are not as complex as the logic circuits are. Therefore, the readiness for changing the device concept and introducing new devices is much more than that of logic circuits. As indications, it is seen that different materials and device types have been applied in radiofrequency electronics, including high-electron-mobility transistors (HEMTs) based on iii–v semiconductors such as GaAs and InP, silicon n-channel MOSFETs and different types of bipolar transistors [23].

Graphene, a new material for transistor channel, was first introduced for application of radiofrequency. It is hoped that by using graphene, which is one-atom-thick layer of graphite, it is possible to fabricate MOSFETs with extremely thin channels, which will make these devices able to be scaled to shorter channel lengths and lower delay without facing the short-channel issues that limits the operating frequency of the current silicon devices. Therefore, proposing new devices would be one of the most promising alternatives to improve silicon [24].

Graphene in its mono-layer form is a pure two-dimensional (2D) material. Its lattice comprises regular hexagons of carbon atoms. The graphene lattice constant, a, is 0.246 nm and the bond length of adjacent carbon atoms, L_b, 0.142 nm. The application of this material has been reported long time ago in [25], when it was not even called graphene. However, all the attempts to make stable graphene all failed. Therefore, for long time it was thought that graphene cannot be existed and stable at room temperature [25]. However, it was experimentally shown to be stable at room temperature in 2004 paper by the Manchester group [26] to start the huge amount of research on this material.

2.6.1 Advantages of Graphene-Based Electronics

In 2004, an extremely high carrier mobility ($\approx$10000 cm^2/V s) of graphene has been experimentally and theoretically shown [26]. However, this property of graphene needs to be discussed in more detail, which is given later in this chapter. Due to its high mobility, if graphene is applied as a material of MOSFETs' channel, those devices could be considered as semi-ballistic transistors. Furthermore, extraordinary high conductance of graphene results in very high current and low delay in carbon-based transistors. The electron or hole transport in graphene occurs in the p-orbitals perpendicular to the surface, and the exceptional transport characteristics have been connected to a single spatially quantized subband populated by donor carriers with low effective mass of $m_e = 0.06 \times m_0$ or by light and heavy holes with masses of $m_h = 0.03 \times m_0$ and $m_h = 0.1 \times m_0$ [24]. Mean free path for carriers of $\lambda \approx 400$ nm at 300 K is another prospect of realizing ballistic devices, even at relaxed feature sizes compared to the state-of-the-art CMOS technology [24].

2.6.2 Disadvantages of Graphene-Based Electronics

In the modern digital circuit, complementary MOS (CMOS) is the dominant technology. A CMOS technology applies both n and p-type FETs in order to make low-power circuits. The main idea is that at final states only one type is on and the other one is completely off so the path between VCC and GND is disconnected.

The major benefit of CMOS over other technologies is that in the final states, a number of the transistors are in off state resulting in having no static current. This feature of silicon MOSFETs makes silicon CMOS enable to offer exceptionally low static power consumption. Consequently, any possible successor to the current MOSFET, which is to be applied in CMOS-like logic circuit should have very good switching characteristic, as well as an I_{on}/I_{off}, in range of 10^4 to 10^7 [22].

To do so, a bandgap of 0.4 eV or more is required in conventional FETs. In addition, to make CMOS circuits, n- and p-type FETs are required with $V_{tn} = -V_{tp}$ for a proper CMOS operation. The major drawback of graphene-based FET is that they are not suitable for CMOS applications. Inferior I_{on}/I_{off} ratios in graphene-based devices due to zero bandgap of unbiased and large-area graphene make inefficient CMOS devices. Conductivity in graphene is at lowest point under 0 V gate bias, but turning off the device is difficult or even impossible at normal temperatures because thermal energy and fluctuations are more than sufficient to produce large carrier populations [1].

Fig. 2.7 Mispositioned CNT resulting in current variation in CNTFETs (extracted from [28])

As a result, leakage current is too high in graphene-based transistors and thus I_{on}/I_{off} ratios become typically just 1 or 2 orders of magnitude, which is not enough for implementing an applicable MOSFET [1]. The next important problem is to find an approach to reliably deposit the nanoribbons in predefined locations for mass and scalable transistor fabrication [1]. Finally, producing scalable and high quality sheets from graphene is an awkward task [1].

Many researchers are working on improving the bandgap in graphene sheet to make it more suitable to be used as channel of low-power transistors. So far, there has been success in providing good semiconducting graphene-based channels using GNR and carbon nanotubes (CNT) resulting in very high-performance transistors.

GNR-based channel opens a bandgap inversely proportional to the width. To gain enough bandgap the width of GNR must be less than 3 nm. However, in that length mobility is degraded. In the case of CNFET circuits, it is costly and very difficult to fabricate them at large scale due to some serious manufacturing issues like variations in doping and diameter of CNTs, unwanted produce of metallic CNTs and mispositioned CNTs shown in 2.7 [27]. As the doping and diameter variations in CNTs result in drain current variations, the major problem is the handling mispositioned and metallic CNTs because they impair the operation of the gate [28] (Fig. 2.7).

2.6.3 Application of Graphene in Electronics

In this section, the potential applications of graphene in digital and analogue electronics are discussed briefly. Several logic gates and arithmetic circuits using graphene and CNT have been proposed in the literature, which are discussed here. In addition, the application of the carbon in analogue devices is introduced.

2.6.3.1 Applications of Graphene in Digital Electronics

Graphene can be used in many applications. For example, it has the potential properties to be suitable component of the next generation of the integrated circuits. In addition, it benefits from an excellent carrier mobility and low noise, allowing that to be applied as the channel of semi-ballistic FETs applying ultra high-speed devices using graphene channel, several high-performance logic and arithmetic

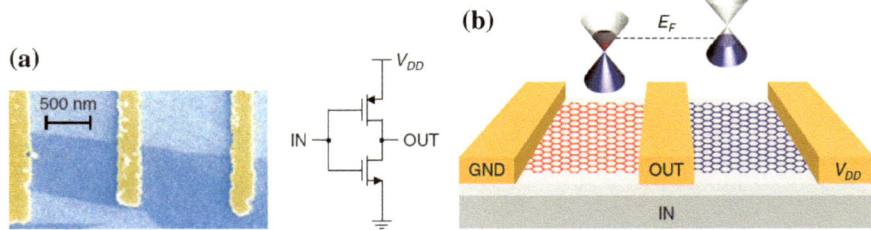

Fig. 2.8 A schematic of the fabricated CMOS inverter (**a**) and Fermi level repositioning in order to implement p-type and n-type FET (**b**) (extracted from [29])

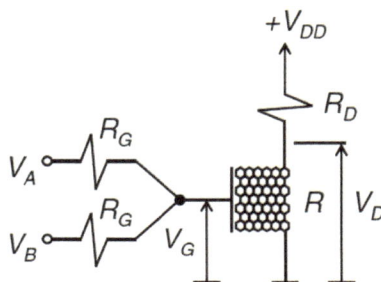

Fig. 2.9 An AND logic gate (**a** and **b**) employing mono-layer graphene transistor. R is the output resistance of the graphene transistor, depending on the voltage of the gate V_G. Obviously this kind of logic dissipates static power but benefits from very low delay using high mobility graphene mono-layer. Since making p-type and n-type channel is still a challenge the superiority of graphene has been verified using undoped channel and RTL logic (extracted from [30])

circuits have been reported. In [29], a complementary inverter using mono-layer graphene nanoribbon is fabricated, and the performance of the circuit is measured. Figures 2.8a, b show the cross section of the fabricated module, and the image of the real fabricated device, respectively.

Reference [30] shows the performance of potential logic gates using mono-layer graphene nanoribbon. Their proposed circuits include several ratio logic gates, such as NAND, NOR, AND, and inverter using graphene. Figure 2.9 shows the schematic of the proposed AND gate. Paper [31] reported high speed full adders using carbon nanotube transistors and capacitors. Very high speed has been resulted compared to the state-of-the-art full adders in the literature.

2.6.3.2   Application of Graphene in Analogue Electronics

An excellent material to be used in high-frequency analogue circuit needs mechanical and thermal stability, high thermal conductivity, superior carrier mobility and extremely high resistance to electro-migration [23]. All these properties have been seen in graphene making that a promising candidate for high-frequency analogue applications. Graphene possessing high carrier speed can present a large small-signal transconductance g_m, determined as dI_{ds}/dV_{gs}, which is the most important parameter in measuring the speed and frequency of a FET and also an amplifier gain. Recently reported results have shown graphene FETs with a cut-off frequency (f_T) of 100 GHz could be fabricated [32]. In addition, it is believed that even graphene FETs with THz frequency could be created using 9 nm graphene channel.

Furthermore, it has been established that graphene shows low level of $1/f$ noise, which is an important criterion for analogue circuits working at high frequencies. Comparing the $1/f$ of graphene with that of the conventional transistors reveals the suitability of graphene in analogue applications in terms of the noise spectral density [32].

2.6.4   Important Graphene Parameters in Connection with FETs

Three important properties of graphene are discussed: the bandgap, carrier transport (mobility and high field transport) at room temperature and 2D nature of graphene.

2.6.4.1   Bandgap Definition in Semiconductors

It is established that large-area graphene behaves like semi-metal and its bandgap is zero. The conduction and valence bands of graphene are cone-shaped and meet each other at the K points of the Brillouin zone (Fig. 2.13b). In logic application, it is needed to switch off the transistors based on the input logic and since the bandgap of large-area graphene is zero, it is not suitable for application of logic circuits. However, it is possible to alter the bandstructure of the graphene and open a bandgap to make a semiconductor. There are three ways to open a bandgap in graphene [33].

1. By making one-dimensional graphene nanoribbons with width less than 10 nm. As the width of the ribbon is reduced the larger bandgap opens.
2. By using two layers of graphene sheets forming bilayer graphene applying different biases to the layers. Increasing the electric field strength causes the bandgap to increase.
3. By applying strain to graphene. See Fig. 2.10 for remarks.

Based on the direction of applying current there are two types of GNR called armchair and zigzag nanoribbon shown in Fig. 2.11. It has been forecast that both

Graphene type	Size	Bandgap	Remarks
SL graphene on SiO_2	LA	No	Experiment and theory
SL graphene on SiO_2	GNR	Yes	Experiment and theory; gap due to lateral confinement*
BL graphene on SiO_2	LA	Yes	Experiment and theory; gap due to symmetry breaking by perpendicular interlayer field
Epitaxial SL	LA	Unknown	Controversial discussion
		Yes	Experiment and theory, gap due to symmetry breaking
		No	Experiment and theory
Epitaxial BL	LA	Yes	Experiment and theory
Epitaxial SL, BL	GNR	Yes	Theory
Strained SL†	LA	Yes	Theory; gap due to level crossing
		No	Theory

Fig. 2.10 Bandgap in graphene. SL: single layer; BL: bilayer; LA: large-area; GNR: graphene nanoribbon. As can be seen existence of bandgap in GNR and BL graphene has been shown by experimental and theoretical studies. However, large-area graphene does not open bandgap (extracted from [1])

(a)   **(b)**

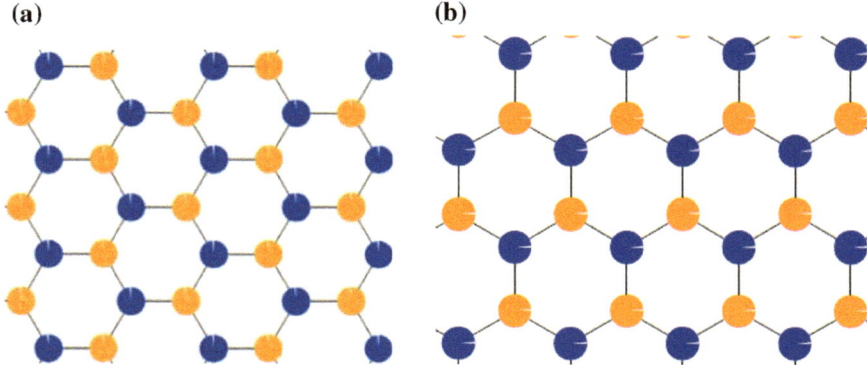

Fig. 2.11 Armchair (**a**) and zigzag (**b**) forms of GNR. The direction of current is assumed from left to right (lateral direction)

armchair and zigzag nanoribbons being ideal types of nanoribbon (Fig. 2.11a, b) open a bandgap, which is appropriately related to the width of the nanoribbon.

Equation 2.7 can be used to calculate the bandgap in armchair GNR, which is called simply GNR in the rest of this thesis.

$$E_g = \frac{2\pi \upsilon_F}{3W} \tag{2.7}$$

The $\hbar$ is reduced Plank's constant, $\upsilon_F = 10^6$ m/s is the Fermi Velocity and W is the GNR's width. In addition, the bandgap in zigzag GNR is calculated from

Fig. 2.12 Bandgap versus
nanoribbon width. As width
of nanoribbon decreases the
bandgap increases [35]

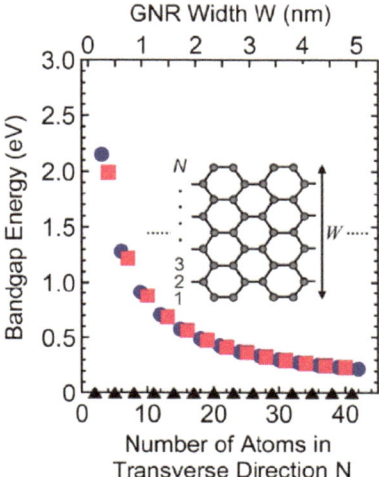

$$E_g = \frac{1.65 \ (eV)}{W_g \ (nm)} \tag{2.8}$$

as reported in [34]. The bandgap opening in GNR has been shown experimentally
for widths down to almost 1 nm [35], and a bandgap of around 0.2 eV has been
demonstrated experimentally and theoretically for widths below 20 nm (Fig. 2.12).
Therefore, to use nanoribbon in graphene transistors, a very narrow sheet of graphene with excellent edges and well-defined width is required to be fabricated. This
is really difficult to fabricate such a material with the available processing equipment of semiconductors. Using unfolding carbon nanotubes, the nanoribbons with
uniform width and optimized edges are produced [1]. However, even a perfect
nanoribbon in terms of edge roughness and width is not completely suitable for
electronics applications.

As the bandgap increases the valence and conductance bands become parabolic
instead of cone-shaped. This causes rise in carrier effective mass, which in turn
reduces the mobility.

In bilayer graphene the conduction and valence bands have parabolic shape near
the k point, but the bandgap is still zero. However, if we apply a suitable electric
field to the bilayer graphene, a bandgap appears and the shape of the bands becomes
so-called Mexican-hat shape (Fig. 2.13b). Theoretical studies show that the bandgap's size is related to strength of the applied electric field. Based on the theoretical
models, it can be as large as 200–250 meV for electric fields with strength around
$1-3 \times 10^7$ V cm^{-1} [36]. There is doubt about the bandgap of large-area mono-layer
graphene. There are reports suggesting the nonzero bandgap of 0.25 eV.

On the other hand, when the material is used in transistor, the transfer characteristic shows no switching-off demonstrating zero bandgap [1]. Finally, applying
strain has been proposed as an approach to open a bandgap in graphene, and the

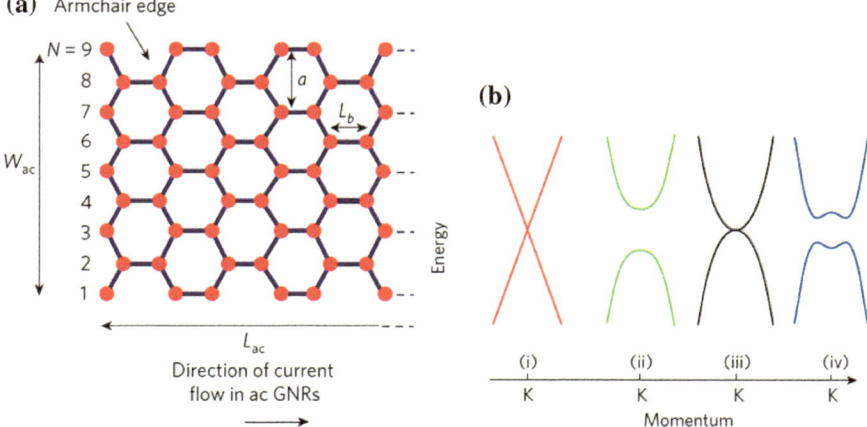

Fig. 2.13 Bandgap related to graphene and GNRs. **a** An armchair GNR (ac-GNR) with length L_{ac} and width W_{ac}. **b** Bandstructure for large-area graphene (i), GNR (ii), bilayer graphene (iii) and bilayer graphene under the influence of an electric field (extracted from [1])

influence of uniaxial strain on the bandstructure has been studied using simulation [37]. Currently, it seems that opening bandgap in graphene using strain is a difficult task in practice since it requires a global uniaxial strain more than 20%. Therefore, although there are some techniques to open a bandgap in graphene, none of them are suitable to be used in realistic applications and industry, since, realizing all those techniques, making nanoribbons, nanotubes and bilayer of graphene, and applying global uniaxial strain more than 20% are difficult and costly due to their need to complicated and accurate machines.

2.6.4.2 Mobility Description in Semiconductors

One of the most promising advantages of graphene is its extraordinary mobility at room temperature. Mobility of exfoliated graphene on SiO_2 has been routinely measured and shown to be 10,000–15,000 cm^2 V^{-1} s^{-1} [1]. Even upper limits of between 40,000 and 70,000 cm^2 V^{-1} s^{-1} have been reported in other studies such as [38]. Moreover, having no of ripples and charged impurities, carrier mobilities of 200,000 cm^2 V^{-1} s^{-1} have been estimated, and a mobility of 10^6 cm^2 V^{-1} s^{-1} was recently suggested for suspended graphene [39].

Finally, the mobility of graphene depends on the surface on which graphene is grown. If the graphene is grown on SiC the mobility is reported to be higher than that of the graphene, which is grown on silicon. The mobility of 5,000 cm^2 V^{-1} s^{-1} has been reported for graphene on SiC [40] while for graphene grown on silicon face the value is 1,000 cm^2 V^{-1} s^{-1} [41]. However, growing graphene on SiC is a

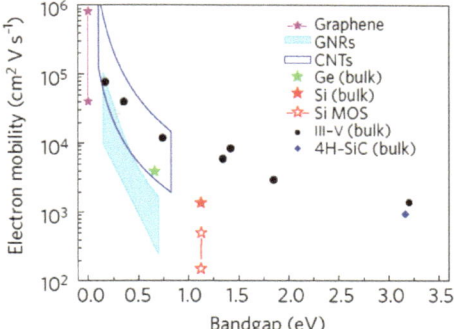

Fig. 2.14 Electron mobility in conventional material such as iii–v compounds InSb, InAs, $In_{0.5}3Ga_{0.47}As$, InP, GaAs, $In_{0.49}Ga_{0.51}P$ and GaN. The values of mobility are related to undoped material except the Si. Moreover, two regions are demonstrated to show the range of mobility in CNT and GNR provided by experimental and theoretical studies. The figure shows that GNR can present lower mobility than that of silicon at bandgap wider than almost 0.5 eV (extracted form [1])

difficult task, which makes silicon face more suitable for electronic applications at the moment. In first graphene FETs, top-gate dielectric affected the mobility [42]. However, the later researches showed that it is possible to make transistors with high mobilities with applying suitable dielectric and deposition process optimization. For example, mobility of around 23,000 cm^2 V^{-1} s^{-1} has been observed by [43].

Although the reported mobilities are impressive, it is important to be careful that they are all related to large-area graphene with zero bandgap, which is not suitable for electronic use. As Fig. 2.14 shows by increasing the bandgap, the mobility decreases. This trend is valid in conventional semiconductors, and it is also predicted to be true in graphene and carbon nanoribbon [44].

If we compare silicon and graphene at the same condition, both having 1.1 eV bandgap, the mobility of graphene is estimated to be less than of silicon. This means that having large bandgap, graphene does not show any advantage over silicon in terms of mobility. The mobilities measured in experiments-less than 200 cm^2 V^{-1} s^{-1} for nanoribbons 1–10 nm wide and 1,500 cm^2 V^{-1} s^{-1} for a nanoribbon with width 14 nm (which is the highest mobility reported so far for a nanoribbon)—agree the theoretical results. Figure 2.15 shows the GNR width respect to mobility.

Therefore, although graphene may provide high-speed operation due to its high carrier mobility, it has the drawback that when it turns on it does not turn off or the ratio of I_{on}/I_{off} is small. Thus its application in logic circuits is not practical for the moment since it results in high power consumption.

Fig. 2.15 As the nanoribbon width is reduced the bandgap gets wider but unfortunately the mobility of graphene decreases from 10^5 cm/Vs for $W = 10$ nm to 10^3 cm/Vs for $W = 2$ nm (extracted from [45])

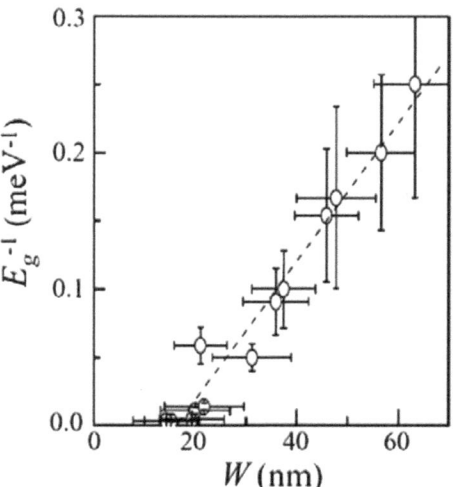

2.6.4.3 High-Field Transport in Semiconducting Devices

In FETs with gates several μm long, carrier's speed is a proper function of mobility. The electric field is not so high in long-channel transistors. However, in nanometre FETs, which the gate length is around few nanometres, the electric field is quite high resulting in velocity saturation. Therefore, the dependence of carrier transport on mobility fades. In order to show this, suppose we have a FET with 100 nm gate length and $V_{ds} = 1$ V. Assuming a voltage drop of 0.3 V along the drain-source resistance, the electric field in the channel is approximately 70 kV cm^{-1}. At that high field, the carrier velocity reaches a saturation point at steady-state, and thus the saturation velocity become an important parameter of carrier transport. In Fig. 2.16 plots of the carrier velocity against the field are illustrated.

For carbon nanotube and the graphene, the highest carrier velocities are predicted to be around 4×10^7 cm s^{-1}, compared to 2×10^7 cm s^{-1} for GaAs and 10^7 cm s^{-1} for silicon. In addition, at high electric fields, the velocity's trend in the nanotube and graphene is not decreasing drastically as it is in the iii–v semiconductors. Unfortunately, at present, no experimental study exists in literature to show the transport in graphene nanoribbons at high electric field and in large-area graphene. But, in some references such as [46] high-field carrier speed of few 10^7 cm s^{-1} in graphene has been reported. Thus, in terms of high-field velocity, graphene and the carbon nanotubes have a little superiority over the silicon devices.

Finally, it has to be said that published mobilities for graphene devices require to be examined in details since the definitions for the mobility of the channel are different in the different papers. Therefore, they cannot be simply compared together. In addition, the methods used to measure the transport characteristics have been only vaguely described in some works. Moreover, the resistance of the drain and source and drain contacts should be eliminated in measuring the transport

Fig. 2.16 Drift velocity of electrons respect to electric field for large-area graphene (extracted from [47])

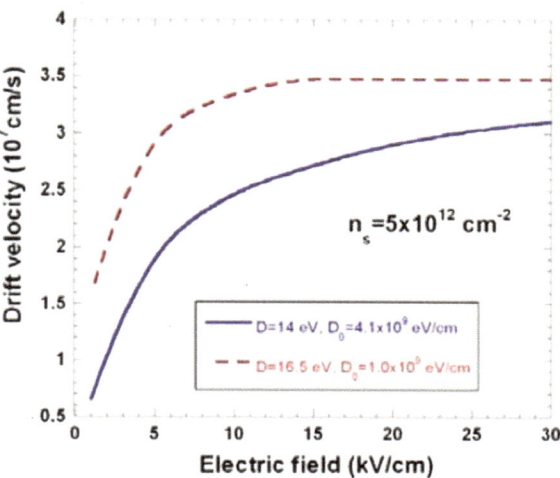

characteristics. However, it is not clear whether this has been done in all the works or not.

However, it is commonly believed that the filed effect mobility of graphene (μ_{FE}) is given by Schwierz [1]

$$\mu_{FE} = \frac{Lg_m}{WC_gV_s},\qquad(2.9)$$

where L, W are channel length and width, respectively, C_g is the gate capacitance, and g_m is transconductance. In addition, there could be different interpretation of capacitance of top-gated GNR FETs. Normally, it is calculated form $C_{ox} = \varepsilon_{ox}/t_{ox}$ defined as the oxide capacitance per unit area, where t_{ox} is the dielectric thickness and ε_{ox} is the dielectric constant of the top-gate. It must be taken into consideration that the quantum capacitance, C_q, is not insignificant value when t_{ox} is small. Therefore, the total gate capacitance should be calculated form $C_G = C_{ox}C_q/(C_{ox} + C_q)$. Particularly, close to the Dirac point, where the current reaches the minimum value, the fieldeffect mobility could be underestimated if the effect of quantum capacitance is ignored.

2.6.4.4 Two-Dimensional Nature of Graphene

Graphene with its 2D nature offers the thinnest possible channels for the FETs. Thus, GNRFETs could be more scalable than the other counterparts are such as Si-based FETs. It is worth to mention that, however, the theory of scaling is valid for semiconducting materials and cannot be applied to semi-metal channels like GNRFETs having zero bandgap. Therefore, scaling theory could be used for graphene nanoribbons with a tunable bandgap but noticeably degraded mobilities than the large-area graphene. As a result, since the high value of mobility is mostly related to large-area graphene, which is not suitable for FETs, it can be said that the most attractive advantage of the graphene is its ability to be extremely scaled down in form of nanoribbon rather than its high mobility.

2.7   Length of Velocity Saturation Region

As discussed before, the length of saturation region can be obtained by solving surface potential at $\phi(L_d) = V_{\text{sat}}$, where V_{sat} is the saturation voltage. Therefore, it is needed to review the surface potential models. Although there are many analytical and semi-analytical models for surface potential of silicon-based devices, there is a lack of research in modelling of this parameter for carbon-based transistors. Therefore, in this section we briefly review the reported models for conventional silicon transistors. In addition, few works focusing on length of saturation velocity region are studied at the end of this section.

Table 2.1 shows important works regarding surface potential modelling. In [14] paper, the modelling starts with

$$\frac{d^2 V}{dx^2} = \frac{qN_{\text{epi}}}{\varepsilon} + \frac{dE_y}{dy}, \tag{2.10}$$

which is a simple one-dimensional model. Using this model the breakdown voltage single-gate power MOSFET is calculated. The results later are recalculated with MEDICI simulator and compared with results computed using the model and they agree well. Imam presented a model for threshold voltage of a typical double-gate MOSFET using a two-dimensional surface potential model. The basic equation is written as

$$\psi(x, y) = \frac{qN_a}{\varepsilon_{\text{si}}} \quad 0 \leq x \leq t_s \text{ and } 0 \leq y \leq L, \tag{2.11}$$

where L and t_s are the channel length and thickness, respectively. Then using separation technique the equation is separated and written as.

Table 2.1 Application of surface potential in modelling of different device characteristics

Approach	Important references
One-dimensional	[2, 14]
Two-dimensional	[9, 49]
Three-dimensional models	[48, 53]

$$\psi(x, y) = V(x) + U(x, y), \tag{2.12}$$

where $V(x)$ is the surface potential at the surface of the channel and $U(x, y)$ is the solution, which accounts for the 2D short-channel effects. Using proper boundary conditions, the potential along the x and y direction can be calculated. A good related review paper was published by Wong, which reviews the physics and models of drain breakdown in short-channel MOSFET. Four mechanisms, namely, (1) avalanche breakdown (MI mode), (2) finite multiplication with positive feedback of the substrate current and (3) parasitic transistor induced breakdown and (4) punch-through, are discussed. The same approach as what Imam used was employed by Yang [9] to calculate breakdown voltage of double-gate power MOSFET.

After the surface potential model is obtained, the lateral electric field is modelled and then using the ionization coefficient of silicon film and avalanche breakdown condition the breakdown voltage is calculated and compared with experimental results. A simple analytical expression of the 3D potential distribution along the channel of lightly doped silicon tri-gate MOSFETs in weak inversion was derived in [48]. It was based on a perimeter-weighted approach of symmetric and asymmetric double-gate MOSFETs. The analytical solution was compared with the numerical solution of the 3D using FlexPDE simulator. Finally using the model, subthreshold slope and short-channel effects are studied. In conclusion, it can be said that surface potential models can be broadly classified into one-dimensional (1D), two-dimensional (2D) and three-dimensional models (3D).

One-dimensional models are simple and reasonably accurate but they are just able to study the surface of the channel. On the other hand, 2D models show more flexibility respect to 1D models and finally, the 3D models can be the most accurate models presented. Differential equations for 1D, 2D and 3D methods are given as

$$\frac{\partial^2 \phi(x)}{\partial^2 x^2} = \frac{qn(x)}{\varepsilon_{channel}} \tag{2.13}$$

$$\frac{\partial^2 \phi(x, y)}{\partial x^2} + \frac{\partial^2 \phi(x, y)}{\partial y^2} = \frac{qn(x, y)}{\varepsilon_{channel}} \tag{2.14}$$

$$\frac{\partial^2 \phi(x, y, z)}{\partial x^2} + \frac{\partial^2 \phi(x, y, z)}{\partial y^2} + \frac{\partial^2 \phi(x, y, z)}{\partial z^2} = \frac{qn(x, y, z)}{\varepsilon_{channel}}, \tag{2.15}$$

where $\varepsilon_{channel}$ is the channel dielectric constant and x, y and z are the dimensions of the channel. In other point of view, surface potential approaches are divided into three categorises.

1. Using separation method
2. Gauss' law
3. Poison equation together with a parabolic function.

In the first approach, it is strongly assumed that potential along the channel can be separated and written as multiplication of two different elements based on the two directions, $\phi(x, y) = \phi(x) + U(x, y)$ and then the resulted equations should be solved for every direction [49]. The resulted partial equations could be solved by applying Fourier series. Calculation can be roughly done by taking approximation and considering only two or three first expressions of the Fourier series [9]. However, the expressions are still complicated if a good agreement is needed between theoretical and simulated results [50].

The second method applies the Gauss law on a rectangular region in the channel [51]. Then special parameters should be proposed in order to get the expression of electric field and surface potential at vertical direction. This method can simply be used in 1D solutions and adequate accuracy could be resulted. In this approach also potential could be modelled only at surface. Typically the modelling starts with

$$\frac{\partial^2 \phi_1(x)}{\partial x^2} = \frac{V_g - V_{bi} - \phi_1(x)}{\lambda^2} = \frac{-q(N+n)}{\varepsilon_g} \tag{2.16}$$

for a single-gate device, for example. Banna [51] used this approach to study the lateral electric field and current of short-channel transistors. Experimental data is used in order to verify the results. To start the modelling application of Guass's law at saturation region is employed. The same approach was used again by Singh [6] for calculation of breakdown of submicron MOSFETs. They reported that a 0.25 um technology single-gate MOSFET experience breakdown at voltages in range of 8–8.7 V. Finally, the third approach employs Poison equation. In this approach, it is assumed that surface potential can be approximately expressed as a parabolic function given as

$$\phi(x, y) = C_0(y) + C_1(y)x + C_2(y)x^2 \tag{2.17}$$

and the coefficients c_0, c_1 and c_2 can be obtained by introducing proper boundary conditions using Gauss law [48]. For interested reader in application of surface potential to extract characteristics of transistors, Table 2.2 is provided here. In addition, the surface potential has been successfully employed in several device structures given in Table 2.3.

To calculate length of velocity saturation region (L_d) from surface potential, it is a common practise to use saturation condition in surface potential equation ($\phi(L_d) = V_{sat}$) [51]. There are models available to calculate length of saturation

Table 2.2 Application of surface potential in modelling of different device characteristics

Device characteristic	Important references
Short-channel effects	[48, 52]
Threshold voltage	[49, 53, 54]
Drain current	[48, 50]
Lateral electric field	[2, 6, 20, 51]
Breakdown voltage (BV)	[2, 6, 20]

Table 2.3 Application of surface potential in modelling of different device structures

Device structure	Important reference papers
Single-gate low-power FETs	[7, 14]
Double-gate transistors	[9]
Tri-gate transistors	[48]
Surrounding-gate devices	[55]
High power transistors	[8, 17]
Silicon nanotransistors	[56]
GNR-based devices	[37]

region [2, 5, 51]. Using the mentioned approach, they normally came up with an expression given as below, for instance, with L_d at both sides of the equation, which should be solved numerically.

A simple equation of L_d proposed in [5] is given as

$$L_d \approx \lambda \ln\left(\frac{a + u + \sqrt{u^2 + 2au + 1}}{a + 1}\right) \tag{2.18}$$

where

$$u = a\left(\cosh\frac{L_d}{\lambda} - 1\right) + \sinh\frac{L_d}{\lambda} \tag{2.19}$$

Since measuring this parameter is quite difficult, analytical modelling is a suitable tool in this case. As a result, there are several models [8, 15, 17, 57, 58] in the literature for conventional devices. Based on empirical results, L_d is a function of oxide thickness, junction depth, bias and channel length. One example is the empirical expression relying on MINIMOS [59] simulation results. However, the proposed model is not well-developed. As an example, some introduced parameters are dependant on device and vary for different devices. Further a more precise model was presented in 1997 by [5], which is given by

$$L_d \approx l_{d0}\ln\left(\frac{a + u + \sqrt{u^2 + 2au + 1}}{a + 1}\right) \tag{2.20}$$

where, $l_{d0} = \sqrt{\varepsilon_{si}t_{ox}x_j/\varepsilon_{ox}}$, $u = (V_D - V_{Dsat})/l_{d0}E_s$ and $a = l_{d0}/(L - 2l_{d0})$. The E_S is defined as the minimum electric field required to secure saturation velocity; V_{ds} and V_{Dsat} are the drain-to-source voltage and the drain saturation voltage of the MOSFET, respectively. The t_{ox} is the thickness of the gate oxide; x_j is the drain junction depth; L is the effective channel length; and ε_{ox} and ε_{Si} are the dielectric constants of gate oxide and silicon, respectively.

Equation 2.20 shows that l_d increases as the drain voltage increases and is governed by channel length. In Fig. 2.17, width of impact ionization region relation is shown respect to the bias and channel length. As numerical results show, the

Fig. 2.17 The effects of drain-source V_{ds} and channel length L on the length of velocity saturation region (Impact ionization length). The L_d increases with V_{ds} and channel length (extracted from [2])

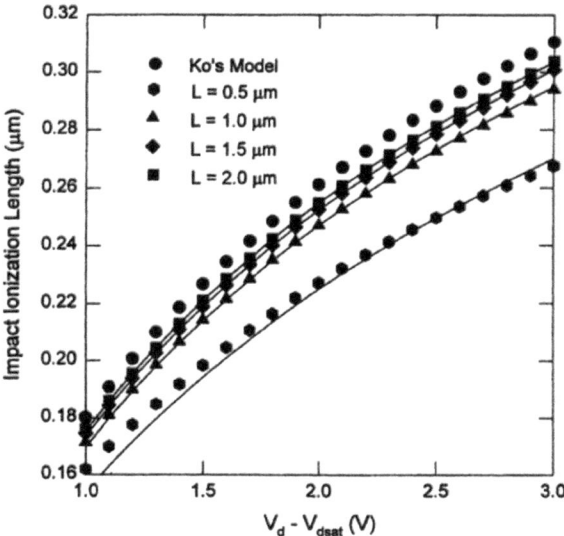

models presented by Eq. 2.20 as L increases, the L_d, which is called impact ionization length, increases too and it is verified well with the compared results. However, the model is only valid for the channel length longer than 500 μm and drain voltage between 1 and 3 V.

2.8 Ionization Coefficient

According to the literature, there has been no attempt to study the ionization coefficient of GNR. However, there are models for ionization coefficient in conventional materials. Table 2.4 introduces the most important works regarding ionization modelling in conventional semiconductors. According to the previously presented methods, calculation of ionization coefficient can be obtained in two different cases. It can be computed at very high electric field, where the Wolf's model [60] is valid or at very low field, where Shockley's model is applicable [61]. However, at moderate values of electric field, none of the mentioned models is valid.

Table 2.4 The most important papers presenting ionization coefficient models in conventional semiconductors

Semiconductor	Important works
Silicon	[11, 12, 60, 61]
InSb	[62, 63]
SiGe	[18]
Silicon with disorders	[10, 64]

Therefore, authors in [65] were motivated to propose a model, which is exact at both low and high electric field values. However, their model is too complicated and time-consuming to be used in real applications [65]. Calculation of breakdown voltage and substrate current is normally includes integration of the ionization coefficient over the length of velocity saturation region [66], therefore, it is necessary to propose a simple mathematical approach for the ionization rate and surface potential model. As a result, for a long time, Shockley's model has been widely applied in modelling the impact ionization and hot electron generation in MOSFETs. According to Shockley's model [61], the impact ionization coefficient, α, which is the number of ionization events per unit length, is determined by the local electric field, F, with the characteristics field strength, B. The impact ionization rate can be approximated by

$$\alpha = A \exp \frac{-B}{F} \tag{2.21}$$

Since expression 2.21 is valid in case of low electric field, it needs two fitting parameters A and B to obtain acceptable values for different processes. They are functions of process, biasing condition and temperature. For silicon surface electrons, A and B are 2.45×10^5 cm^{-1} and 1.92×10^6 V/cm, respectively [61]. The previous work done for modelling this parameter in silicon is reviewed in [12]. We broadly categorize the models presented in the previous works into three types called, equilibrium state (ES), lucky ballistic (LB) and lucky drift (LD) models.

The first attempt to calculate ionization coefficient in semiconductors was made by Wolff [60], which is called ES model. In the ES model, it is shown that the electrons in the tail of the equilibrium distribution can be regarded energetic enough to create an electron–hole pair [11]. The general form of the ES-based models is known as [11, 12].

$$\alpha \approx \exp \frac{-a0}{F^2} \tag{2.22}$$

where F is the electric field strength and $a0$ is a constant. A principal objections to this model is that it is possible that impact ionization may be more associated with non-equilibrium electrons than with those in a nearly isotropic equilibrated distribution. This objection formed the basis of a quite different approach by Shockley [61] LB models, which has become known as the lucky-electron model. In the LB model, impact ionization is shown to be produced by electrons, which happened to avoid collision [11]. If λ_m, which is the momentum relaxation mean free path, is considered as a constant, the probability of an electron avoiding a collision is [11]

$$P_a = \exp\left(\frac{-E_t}{qF\lambda_m}\right) \tag{2.23}$$

and so the ionization coefficient, in the simplest formulation, takes the form [11]

$$\alpha = \frac{1}{l_0}\exp\left(\frac{-E_t}{qF\lambda_m}\right), \tag{2.24}$$

where E_t is the ionization energy, q is the charge magnitude and $l0 = E_t/qF$. The electric field dependence of LB models are widely different from what was resulted by Wolff's models. The dependence in LB models is much more than that of ES models. In practice, it is hard to decide accurately between LB and ES approaches since the range, where α is calculated is too small and both of them have been widely used in the literature. The most important objection against LB approach is that the collision path for carriers is expected to be around 5 nm, which is too short to result lucky-electron with sufficient probability.

As a result, LD model is proposed. In this model, the theory is that carriers could drift in an electric field having momentum-relaxing collisions, a determined drift velocity, and no noticeable energy-relaxing collision at the same period [11]. This state is called lucky drift [11, 64].

Considering λ_m and λ_E to be independent of energy, the ionization coefficient based on LD model is given by Rubel et al. [10]

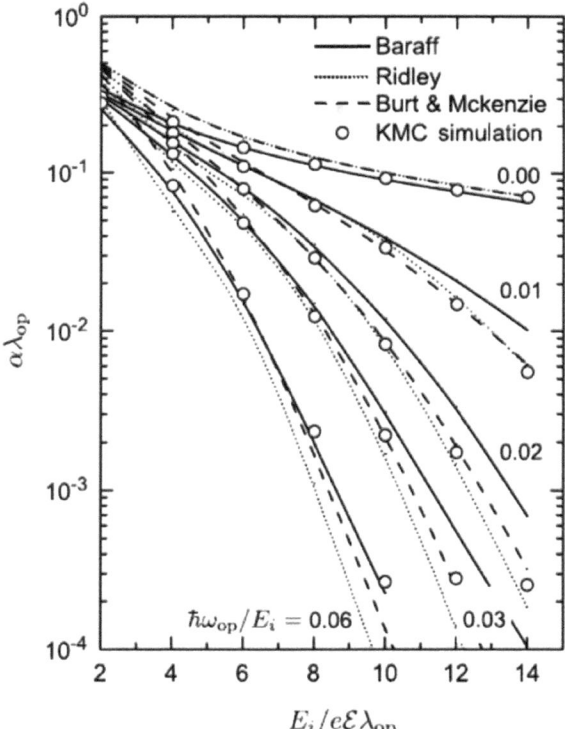

Fig. 2.18 Ionization coefficient versus electric field (ε) and ionization threshold energy (E_i) employing different approaches (extracted from [10])

$$\alpha = \frac{1}{\lambda_m} \frac{r\left(\exp\left(\frac{-l_0}{\lambda_E}\right) - r\exp\left(\frac{-l_0}{\lambda_m}\right)\right)}{1 - \exp\left(\frac{-l_0}{\lambda_E}\right) - r^2\left(1 - \exp\left(\frac{-l_0}{\lambda_m}\right)\right)}, \tag{2.25}$$

where λ_E is the energy relaxation mean free path, and $r = \lambda_m/\lambda_E$. The last approach so far produced the most reliable results in literature. Figure 2.18 shows the results of different modelling approaches.

2.9 Introduction on Graphene

In this section, important carbon-related works are briefly reviewed, which are divided into two subsections. The first subsection, reviews the most important experimental and theoretical studies on carbon-based electronics and next subsection, introduced the most significant works on graphene-based transistors.

2.9.1 Experimental Works and Analytical Models Related to Carbon-Based Electronics

The properties of CNTs have been modelled and studied through experimental and analytical attempts. Since the scope of this research is limited to breakdown voltage, other works only summarized in form of tables for interested readers. Table 2.5 introduces the most important review papers regarding graphene and GNRFETs, which is recommended for any researcher in this field. In addition, it acquaints significant works in connection with other properties and applications in graphene. Among the numerous analytical models presented for graphene's properites and its current characteristic, there are few circuit-level models such as [67, 68], which can simply be used in circuit simulators such as HSPICE. One of the most accurate proposed models is Deng's CNT model, which considers non-idealizes and parasitic capacitances [67].

Table 2.5 Important review papers and other works related to various topics on graphene

Topic	Important references
Review papers	[1, 39, 69]
Fabrication methods	[39, 46, 57, 70–74]
Doping approaches	[75]
Design methodology	[76]
Arithmetic circuits	[31, 77]
Graphene logic gates	[29, 30]
Graphene transistors	[1, 37, 42, 78–80]

Table 2.6 Important models proposed for different properties of graphene

Property	Important references
Bandgap	[33, 35]
Conductance	[81]
Mobility and velocity	[47, 70, 74, 82, 83]
Quantum and classic capacitance	[84, 85]
Fermi velocity	[86]
Current	[37, 78, 82, 87, 88]
Scattering	[46, 47, 72, 74, 82]

The paper presents a circuit-compatible compact model for the intrinsic channel region of the MOSFET-like single-walled carbon nanotube field-effect transistors (CNFETs). Their model for CNFET is valid for a wide range of diameters and chiralities. In addition, it covers CNFET with either semiconducting carbon nanotube (CNT) conducting or metallic channel [67].

As graphene showed properties that are more promising compared to CNT, huge amount of interest has been drawn into modelling of this material. As an example, the mispositioned CNTs and metallic CNTs are no more issues [28]. In addition, the fabrication process is simpler than that of CNTs. As a result, many analytical and semi-analytical models are available for graphene specially graphene nanoribbon, which categorized in Table 2.6.

Beside the theoretical studies on graphene, many researchers are working on fabricating of graphene-based transistors. So far, samples with the highest quality has produced by the original mechanical exfoliation, but the method is not suitable for mass production because it is neither high-yield nor high throughput meaning that this approach is time-consuming and the yield is too low to be used in industry. Yield is is calculated from,

$$\text{yeild} = \left(\frac{\text{defect per unit area} \times \text{die area}}{3} \right)^{-3} \tag{2.26}$$

There are alternatives to mechanical exfoliation including four approaches, which are listed as below [69].

1. Mechanical exfoliation
2. Attempts to catalyze growth in situ on a substrate
3. Bottom-up methods to grow graphene directly from organic precursors
4. Chemical efforts to exfoliated and stabilize individual sheets in solution.

None of these approaches are satisfying too for mass production. In case of producing graphene, chemically, perfect exfoliation in solution needs the 2D crystal to be modified extensively, which results in device performance degradation. Uniform and large-area single layer can alternatively be produced by bottom-up techniques. Due to side reactions and insoluble macromolecules, organic synthesize of the graphene is a size limited process. There are two other approaches, growth of monolayers on substrate by using chemical vapour deposition or CVD and silicon carbide reduction. Careful controlling of the conditions is needed after nucleating a sheet in order to promote crystal growth and avoid seeding the second layer or making grain boundaries. To put it in nutshell, although tremendous progress has been made in fabrication process, mechanical exfoliation using cellophane tape is still the most highest quality approach to produce graphene flakes for small scale production.

In terms of device concept, several devices have been already proposed and experimentally examined. Top-gated devices, which use only one gate on top and double-gate devices, which use top and back gates, for better controlling the electric field and surface potential. Other researchers tried to improve the bandgap by increasing the number of nanoribbons. These devices are called mono-layer, bilayer, tri-layer and multilayer GNR transistors based on the number of nanoribbon layers used.

2.9.2 Review of the Most Important Graphene-Based Transistors

Implementing the graphene MOS transistor was one of the most important results reported in 2004, which was presented by the Manchester group [26]. The nature of the proposed device was a SiO_2 layer with thickness of 300 nm below the graphene. This layer designed to act as back gate dielectric and the back gate was made by a layer of doped silicon (Fig. 2.19a)

The proposed device was acceptable to proof the concept. However, it suffered from significant parasitic capacitances due to use of back gate making the concept impossible to be integrated with the other devices. As a result, to use graphene transistor in real situations, a top-gate is required. In 2007, the first top-gated graphene transistor was proposed in [42], which was a very important milestone in implementing graphene-based transistors and the research was accelerated after that in this field (Fig. 2.19b).

Three methods have been reported to make top-gated graphene.

1. Using exfoliated graphene [42].
2. Growing on metals. (i.e. example copper and nickel [89]).
3. Using epitaxial graphene with top-gate dielectric of SiO_2, Al_2O_3 and HfO_2 [90].

Large-area graphene have been used to form the channel of the top-gated graphene FETs.

As we know, the large-area graphene does not open bandgap, therefore, when the transistor is turned on, it does not switch off.

As Fig. 2.20a shows transistors made by a large-area graphene are unique in terms of the current-voltage characteristic. The potential difference between the gates (top and back gates) and the channel controls the carrier density and type in

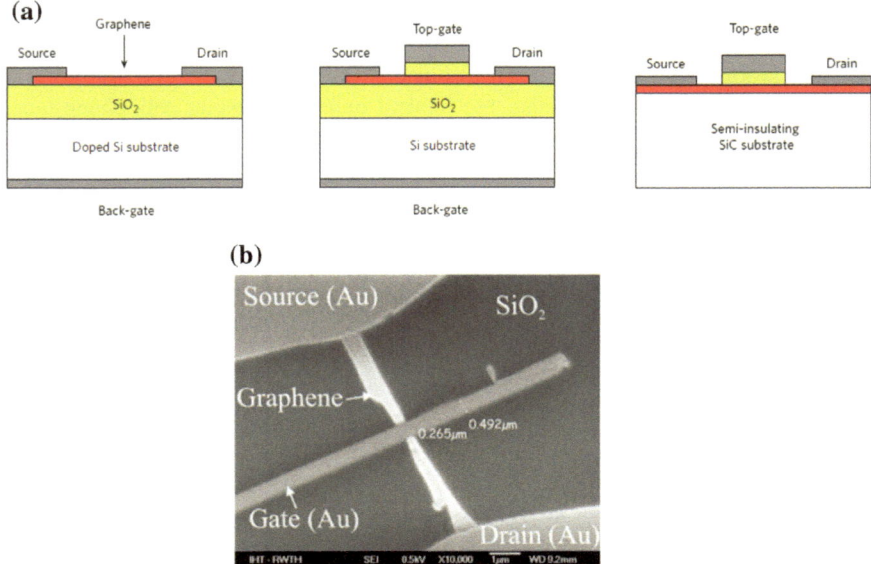

Fig. 2.19 a Several structures to make transistors based on graphene. From left, back-gated, double-gated employing exfoliated graphene channel or graphene, which has been grown on metal, and top-gated with a channel of epitaxial graphene [1]

the channel. If a positive voltage is applied the number of electrons is increased in the channel and therefore, an n-type channel is formed. Otherwise, if a negative voltage is used, a p-type channel is formed due to increase in the density of holes in the channel. As a result, these transistors can be both n- and p-types depending the type potential applied (negative or positive), which is separated by the Dirac point as shown in Fig. 2.20a.

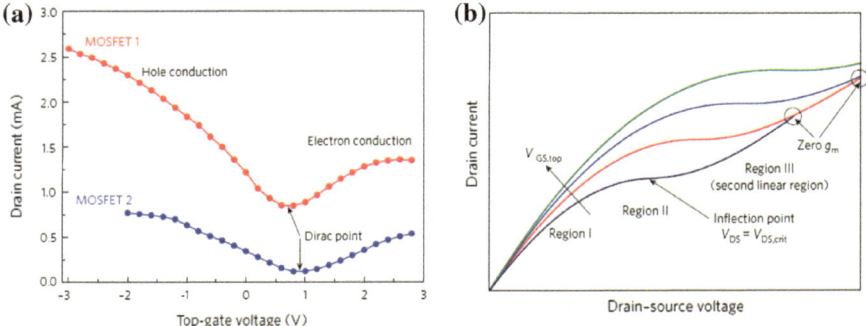

Fig. 2.20 Drain current against drain-source voltage in large-area graphene-based FET. **a** Transfer characteristics consists of two linear regions and one saturation region. **b** Transfer characteristics cross at high values of V_{ds} meaning that the gate cannot control the current (extracted from [90])

Positioning of the Dirac point in these transistors can be performed by adjusting the following approaches [1].

1. The difference between the work functions of the gate and the graphene
2. The charges' density and type at the interface points between channel and top- and back-gates
3. Graphene's doping.

So far the published results for the ratio of I_{on}/I_{off} have been in range of 2–20. In terms of device speed, the graphene has a drawback over conventional devices. Because graphene-based transistors show no saturation or only a week saturation in their transfer characteristic [91]. However, unusual kind of saturation has been seen in some graphene transistors (Fig. 2.20b). As shown in Fig. 2.20b in some conditions, there are two linear regions separated by a saturation region [90].

Based on the reasoning presented in [1], these transistors work as follows. If the V_{ds} is small, the entire channel is n-type and the transistor functions in a linear-like region (region I). By rising the drain-source voltage, current starts to reach a saturation point until the inflection point, on which $V_{ds} = V_{ds,crit}$, is reached (region II). If the V_{ds} increases and exceeds the $V_{ds,crit}$, then the channel is turned from n-type to p-type and therefore, the conduction enters the second linear region. (Region II in Fig. 2.20b).

As the channel in graphene-based transistors have almost zero bandgap, the transfer characteristics overlap at high values of V_{ds}. This leads to a zero or negative transconductance, which is a very undesirable property.

Beside the issues regarding opening a bandgap in graphene nanoribbons, there are other issues in using graphene as transistor channel in digital applications. First, a significantly thick oxide is needed in fabrication of these devices. As a result, relatively high voltage is required to switch the transistor on. While almost the same device in silicon needs only around 1 V to be turned on.

In addition, to form a functional CMOS logic, both n-type and p-type devices with suitable and threshold voltages are necessary. While such devices have not been reported so far GNRFETs with top-gate have been recently reported in [43]. Using a thin top-gate dielectric made from high-k dielectric of HfO_2, the device shows high on/off current ratio of around 70 at room temperature and excellent transconductance of almost 3.2 mS μm^{-1}, which is higher than that of the most state-of-the-art similar silicon devices and iii–v HEMTs. Another way to open a bandgap in graphene is to use two layers of graphene called bilayer graphene and use different bias voltages on the layers. Bilayer graphene transistor was examined by simulation and experiment [92]. The I_{on}/I_{off} ratio has been reported to be 2000 at low and 100 at room temperature. Although it is still not enough for digital applications, it is significant improvement, more than 10 times, over the devices using large-area graphene.

References

1. F. Schwierz, Graphene transistors. Nat. Nanotechnol. **5**(7), 487–496 (2010)
2. H. Wong, Drain breakdown in submicron MOSFETs: a review. Microelectron. Reliab. **40**(1), 3–15 (2000)
3. N.D. Arora, M.S. Sharma, MOSFET substrate current model for circuit simulation. IEEE Trans. Electron Devices **38**(6), 1392–1398 (1991)
4. G. Gildenblat, X. Li, W. Wu, H. Wang, A. Jha, R. Van Langevelde, G.D. Smit, A.J. Scholten, D. Klaassen, PSP: an advanced surface-potential-based MOSFET model for circuit simulation. IEEE Trans. Electron Devices **53**(9), 1979–1993 (2006)
5. H. Wong, M. Poon, Approximation of the length of velocity saturation region in MOSFET's. IEEE Trans. Electron Devices **44**(11), 2033–2036 (1997)
6. A. Singh, Study of avalanche breakdown (MI) mode in sub micron MOSFET device. Microelectron. Int. **22**(1), 16–20 (2005)
7. B.-K. Kim, E.-K. Jeon, J.-J. Kim, J.-O. Lee, Positioning of the Fermi level in graphene devices with asymmetric metal electrodes. J. Nanomaterials **2010**, 8 (2010)
8. L.M. Dang, Drain-voltage dependence of IGFET turn-on voltage. Solid-State Electron. **20** (10), 825–830 (1977)
9. W. Yang, X. Cheng, Y. Yu, Z. Song, D. Shen, A novel analytical model for the breakdown voltage of thin-film SOI power MOSFETs. Solid-state Electron. **49**(1), 43–48 (2005)
10. O. Rubel, A. Potvin, D. Laughton, Generalized lucky-drift model for impact ionization in semiconductors with disorder. J. Phys. Condens. Matter **23**(5), 055802 (2011)
11. B. Ridley, Lucky-drift mechanism for impact ionisation in semiconductors. J. Phys. C: Solid State Phys. **16**(17), 3373 (1983)
12. W. Maes, K. De Meyer, R. Van Overstraeten, Impact ionization in silicon: a review and update. Solid-State Electron. **33**(6), 705–718 (1990)
13. V. Su, I. Lin, J. Kuo, G. Lin, D. Chen, C. Yeh, C. Tsai, M. Ma, Breakdown behavior of 40-nm PD-SOI NMOS device considering STI-induced mechanical stress effect. IEEE Electron Device Lett. **29**(6), 612–614 (2008)
14. D. Križaj, G. Charitat, S. Amon, A new analytical model for determination of breakdown voltage of Resurf structures. Solid-State Electron. **39**(9), 1353–1358 (1996)
15. I.-J. Kim, S. Matsumoto, T. Sakai, T. Yachi, Analytical approach to breakdown voltages in thin-film SOI power MOSFETs. Solid-State Electron. **39**(1), 95–100 (1996)
16. B.J. Baliga, *Fundamentals of Power Semiconductor Devices* (Springer, Berlin, 2010)
17. W. Fulop, Calculation of avalanche breakdown voltages of silicon p-n junctions. Solid-State Electron. **10**(1), 39–43 (1967)
18. K. Yeom, J. Hinckley, J. Singh, Calculation of electron and hole impact ionization coefficients in SiGe alloys. J. Appl. Phys. **80**(12), 6773–6782 (1996)
19. W. Krautschneider, A. Kohlhase, H. Terletzki, Scaling down and reliability problems of gigabit CMOS circuits. Microelectron. Reliab. **37**(1), 19–37 (1997)
20. A. Shah, P. Yang, MOS technology: trends and challenges in the ULSI era, in *20th International Conference on Microelectronics, 1995. Proceedings* (1995), pp. 3–10
21. M. Bohr, A 30 year retrospective on Dennard's MOSFET scaling paper. IEEE Solid-State Circ. Soc. Newslett. **12**(1), 11–13 (2007)
22. I. Aberg, J.L. Hoyt, Hole transport in UTB MOSFETs in strained-Si directly on insulator with strained-Si thickness less than 5 nm. IEEE Electron Device Letters **26**(9), 661–663 (2005)
23. G.E. Moore, "No exponential is forever: but" Forever" can be delayed![semiconductor industry]", in *IEEE International Solid-State Circuits Conference, 2003. Digest of Technical Papers. ISSCC* (2003), pp. 20–23
24. E. Takeda, Challenges for giga-scale integration. Microelectron. Reliab. **37**(7), 985–1001 (1997)
25. J.W. May, Platinum surface LEED rings. Surf. Sci. **17**(1), 267–270 (1969)

26. K.S. Novoselov, A.K. Geim, S. Morozov, D. Jiang, Y. Zhang, S.A. Dubonos, I. Grigorieva, A. Firsov, Electric field effect in atomically thin carbon films. Science **306**(5696), 666–669 (2004)
27. S. Bobba, J. Zhang, A. Pullini, D. Atienza, G. De Micheli, Design of compact imperfection-immune CNFET layouts for standard-cell-based logic synthesis, in *Proceedings of the Conference on Design, Automation and Test in Europe* (2009), pp. 616–621
28. N. Patil, J. Deng, A. Lin, H.-S.P. Wong, S. Mitra, Design methods for misaligned and mispositioned carbon-nanotube immune circuits. IEEE Trans. Comput. Aided Des. Integr. Circ. Syst. **27**(10), 1725–1736 (2008)
29. F. Traversi, V. Russo, R. Sordan, Integrated complementary graphene inverter. (2009) arXiv preprint:arXiv:0904.2745
30. R. Sordan, F. Traversi, V. Russo, Logic gates with a single graphene transistor. Appl. Phys. Lett. **94**(7), 073305 (2009)
31. K. Navi, M. Rashtian, A. Khatir, P. Keshavarzian, O. Hashemipour, High speed capacitor-inverter based carbon nanotube full adder. Nanoscale Res. Lett. **5**(5), 859–862 (2010)
32. X. Yang, G. Liu, A.A. Balandin, K. Mohanram, Triple-mode single-transistor graphene amplifier and its applications. ACS Nano **4**(10), 5532–5538 (2010)
33. E. Sano, T. Otsuji, Bandgap engineering of bilayer graphene for field-effect transistor channels. Jpn. J. Appl. Phys. **48**(9R), 091605 (2009)
34. H. Raza, Zigzag graphene nanoribbons: bandgap and midgap state modulation. J. Phys. Condens. Matter **23**(38), 382203 (2011)
35. M.Y. Han, B. Özyilmaz, Y. Zhang, P. Kim, Energy band-gap engineering of graphene nanoribbons. Phys. Rev. Lett. **98**(20), 206805 (2007)
36. P. Gava, M. Lazzeri, A.M. Saitta, F. Mauri, Ab initio study of gap opening and screening effects in gated bilayer graphene. Phys. Rev. B **79**(16), 165431 (2009)
37. Q. Zhang, T. Fang, H. Xing, A. Seabaugh, D. Jena, Graphene nanoribbon tunnel transistors. IEEE Electron Device Lett. **29**(12), 1344–1346 (2008)
38. J.-H. Chen, C. Jang, S. Xiao, M. Ishigami, M.S. Fuhrer, Intrinsic and extrinsic performance limits of graphene devices on SiO$_2$. Nat. Nanotechnol. **3**(4), 206–209 (2008)
39. A.K. Geim, K.S. Novoselov, The rise of graphene. Nat. Mater. **6**(3), 183–191 (2007)
40. J. Kedzierski, P.-L. Hsu, P. Healey, P.W. Wyatt, C.L. Keast, M. Sprinkle, C. Berger, W.A. De Heer, Epitaxial graphene transistors on SiC substrates. IEEE Trans. Electron Devices **55**(8), 2078–2085 (2008)
41. C. Berger, Z. Song, X. Li, X. Wu, N. Brown, C. Naud, D. Mayou, T. Li, J. Hass, A.N. Marchenkov, Electronic confinement and coherence in patterned epitaxial graphene. Science **312**(5777), 1191–1196 (2006)
42. M.C. Lemme, T.J. Echtermeyer, M. Baus, H. Kurz, A graphene field-effect device. (2007) arXiv preprint cond-mat/0703208
43. L. Liao, J. Bai, Y. Qu, Y.-C. Lin, Y. Li, Y. Huang, X. Duan, High-κ oxide nanoribbons as gate dielectrics for high mobility top-gated graphene transistors. Proc. Natl. Acad. Sci. **107**(15), 6711–6715 (2010)
44. V. Perebeinos, J. Tersoff, P. Avouris, Electron-phonon interaction and transport in semiconducting carbon nanotubes. Phys. Rev. Lett. **94**(8), 086802 (2005)
45. M. Bresciani, P. Palestri, D. Esseni, L. Selmi, Simple and efficient modeling of the E–k relationship and low-field mobility in Graphene Nano-Ribbons. Solid-State Electron. **54**(9), 1015–1021 (2010)
46. E. Hwang, S.D. Sarma, Acoustic phonon scattering limited carrier mobility in two-dimensional extrinsic graphene. Phys. Rev. B **77**(11), 115449 (2008)
47. R. Shishir, D. Ferry, Velocity saturation in intrinsic graphene. J. Phys. Condens. Matter **21**(34), 344201 (2009)

48. A. Tsormpatzoglou, C. Dimitriadis, R. Clerc, G. Pananakakis, G. Ghibaudo, Semianalytical modeling of short-channel effects in lightly doped silicon trigate MOSFETs. IEEE Trans. Electron Devices **55**(10), 2623–2631 (2008)
49. M.A. Imam, M.A. Osman, A.A. Osman, Threshold voltage model for deep-submicron fully depleted SOI MOSFETs with back gate substrate induced surface potential effects. Microelectron. Reliab. **39**(4), 487–495 (1999)
50. A. Tsormpatzoglou, C. Dimitriadis, R. Clerc, Q. Rafhay, G. Pananakakis, G. Ghibaudo, Semi-analytical modeling of short-channel effects in Si and Ge symmetrical double-gate MOSFETs. IEEE Trans. Electron Devices **54**(8), 1943–1952 (2007)
51. M. El Banna, M. El Nokali, A pseudo-two-dimensional analysis of short channel MOSFETs. Solid-state Electron. **31**(2), 269–274 (1988)
52. K.K. Young, Short-channel effect in fully depleted SOI MOSFETs. Electron Devices, IEEE Transactions on **36**(2), 399–402 (1989)
53. G. Katti, N. DasGupta, A. DasGupta, Threshold Voltage model for mesa-isolated small geometry fully depleted SOI MOSFETs based on analytical solution of 3-D Poisson's equation. IEEE Trans. Electron Devices **51**(7), 1169–1177 (2004)
54. D. Munteanu, J. Autran, S. Harrison, Quantum short-channel compact model for the threshold voltage in double-gate MOSFETs with high-permittivity gate dielectrics. J. Non-Cryst. Solids **351**(21), 1911–1918 (2005)
55. N. Collaert, A. Dixit, K. Anil, R. Rooyackers, A. Veloso, K. De Meyer, Shift and ratio method revisited: extraction of the fin width in multi-gate devices. Solid-state Electron. **49**(5), 763–768 (2005)
56. D. Munteanu, J.-L. Autran, X. Loussier, S. Harrison, R. Cerutti, T. Skotnicki, Quantum short-channel compact modelling of drain-current in double-gate MOSFET. Solid-state Electron. **50**(4), 680–686 (2006)
57. F. Fang, A. Fowler, Hot electron effects and saturation velocities in silicon inversion layers. J. Appl. Phys. **41**(4), 1825–1831 (1970)
58. W. Lu, C.M. Lieber, Nanoelectronics from the bottom up. Nat. Mater. **6**(11), 841–850 (2007)
59. A. Schütz, S. Selberherr, H.W. Pötzl, A two-dimensional model of the avalanche effects in MOS transistors. Solid-State Electron. **25**(3), 177–183 (1982)
60. P. Wolff, Theory of electron multiplication in silicon and germanium. Phys. Rev. **95**(6), 1415 (1954)
61. W. Shockley, Problems related top-n junctions in silicon. Cechoslovackij fiziceskij zurnal B **11**(2), 81–121 (1961)
62. J. Devreese, R. van Welzenis, R. Evrard, Impact ionisation probability in InSb. Appl. Phys. A **29**(3), 125–132 (1982)
63. J. Hong, T. Kim, S. Joo, J.D. Song, S.H. Han, K.-H. Shin, J. Chang, Magnetic field dependent impact ionization in InSb. (2012) arXiv preprint arXiv:1206.1094
64. K. Jandieri, O. Rubel, S. Baranovskii, A. Reznik, J. Rowlands, S. Kasap, Lucky-drift model for impact ionization in amorphous semiconductors. J. Mater. Sci. Mater. Electron. **20**(1), 221–225 (2009)
65. Y. Okuto, C. Crowell, Energy-conservation considerations in the characterization of impact ionization in semiconductors. Phys. Rev. B **6**(8), 3076 (1972)
66. Y. El-Mansy, A. Boothroyd, A simple two-dimensional model for IGFET operation in the saturation region. IEEE Trans. Electron Devices **24**(3), 254–262 (1977)
67. J. Deng, H.P. Wong, A compact SPICE model for carbon-nanotube field-effect transistors including nonidealities and its application—Part I: model of the intrinsic channel region. IEEE Trans. Electron Devices **54**(12), 3186–3194 (2007)
68. S. Sinha, A. Balijepalli, Y. Cao, Compact model of carbon nanotube transistor and interconnect. IEEE Trans. Electron Devices **56**(10), 2232–2242 (2009)
69. M.J. Allen, V.C. Tung, R.B. Kaner, Honeycomb carbon: a review of graphene. Chem. Rev. **110**(1), 132–145 (2009)
70. V.E. Dorgan, M.-H. Bae, E. Pop, Mobility and saturation velocity in graphene on SiO_2. (2010). arXiv preprint arXiv:1005.2711

71. E.V. Castro, K. Novoselov, S. Morozov, N. Peres, J.L. dos Santos, J. Nilsson, F. Guinea, A. Geim, A.C. Neto, Electronic properties of a biased graphene bilayer. J. Phys. Condens. Matter **22**(17), 175503 (2010)
72. W.-K. Tse, E. Hwang, S.D. Sarma, Ballistic hot electron transport in graphene. Appl. Phys. Lett. **93**(2), 023128 (2008)
73. I. Meric, M.Y. Han, A.F. Young, B. Ozyilmaz, P. Kim, K.L. Shepard, Current saturation in zero-bandgap, top-gated graphene field-effect transistors. Nat. Nanotechnol. **3**(11), 654–659 (2008)
74. T. Fang, A. Konar, H. Xing, D. Jena, Mobility in semiconducting graphene nanoribbons: phonon, impurity, and edge roughness scattering. Phys. Rev. B **78**(20), 205403 (2008)
75. G. Giovannetti, P. Khomyakov, G. Brocks, V. Karpan, J. Van den Brink, P. Kelly, Doping graphene with metal contacts. Phys. Rev. Lett. **101**(2), 026803 (2008)
76. W.J. Yu, U.J. Kim, B.R. Kang, I.H. Lee, E.-H. Lee, Y.H. Lee, Adaptive logic circuits with doping-free ambipolar carbon nanotube transistors. Nano Lett. **9**(4), 1401–1405 (2009)
77. M. Ghadiry, A.A. Manaf, M.T. Ahmadi, H. Sadeghi, M.N. Senejani, Design and analysis of a new carbon nanotube full adder cell. J. Nanomaterials **2011**, 36 (2011)
78. M. Cheli, G. Fiori, G. Iannaccone, A semianalytical model of bilayer-graphene field-effect transistor. IEEE Trans. Electron Devices **56**(12), 2979–2986 (2009)
79. G. Fiori, G. Iannaccone, Ultralow-voltage bilayer graphene tunnel FET. IEEE Electron Device Lett. **30**(10), 1096–1098 (2009)
80. G. Liang, N. Neophytou, D.E. Nikonov, M.S. Lundstrom, Performance projections for ballistic graphene nanoribbon field-effect transistors. IEEE Trans. Electron Devices **54**(4), 677–682 (2007)
81. M.T. Ahmadi, Z. Johari, N.A. Amin, A.H. Fallahpour, R. Ismail, Graphene nanoribbon conductance model in parabolic band structure. J. Nanomaterials **2010**, 12 (2010)
82. D. Berdebes, T. Low, M. Lundstrom, B.N. Center, Low bias transport in graphene: an introduction. (2009)
83. N. Peres, The transport properties of graphene. J. Phys. Condensed Matter: Inst. Phys. J. **21**(32), 323201–323201 (2009)
84. A. Petrovskaya, V. Zubkov, Capacitance-voltage study of heterostructures with InGaAs/GaAs quantum wells in the temperature range from 10 to 320 K. Semiconductors **43**(10), 1328–1333 (2009)
85. A. Shylau, J. Kłos, I. Zozoulenko, Capacitance of graphene nanoribbons. Phys. Rev. B **80**(20), 205402 (2009)
86. G. Borghi, M. Polini, R. Asgari, A. MacDonald, Fermi velocity enhancement in monolayer and bilayer graphene. Solid State Commun. **149**(27), 1117–1122 (2009)
87. M. Dragoman, D. Dragoman, Graphene-based quantum electronics. Prog. Quantum Electron. **33**(6), 165–214 (2009)
88. M. Cheli, P. Michetti, G. Iannaccone, Model and performance evaluation of field-effect transistors based on epitaxial graphene on SiC. IEEE Trans. Electron Devices **57**(8), 1936–1941 (2010)
89. X. Li, W. Cai, J. An, S. Kim, J. Nah, D. Yang, R. Piner, A. Velamakanni, I. Jung, E. Tutuc, Large-area synthesis of high-quality and uniform graphene films on copper foils. Science **324**(5932), 1312–1314 (2009)
90. J. Moon, D. Curtis, M. Hu, D. Wong, C. McGuire, P. Campbell, G. Jernigan, J. Tedesco, B. VanMil, R. Myers-Ward, Epitaxial-graphene RF field-effect transistors on Si-face 6H-SiC substrates. IEEE Electron Device Lett. **30**(6), 650–652 (2009)
91. Y.-M. Lin, K.A. Jenkins, A. Valdes-Garcia, J.P. Small, D.B. Farmer, P. Avouris, Operation of graphene transistors at gigahertz frequencies. Nano Lett. **9**(1), 422–426 (2008)
92. F. Xia, D.B. Farmer, Y.-M. Lin, P. Avouris, Graphene field-effect transistors with high on/off current ratio and large transport band gap at room temperature. Nano Lett. **10**(2), 715–718 (2010)

Chapter 3
Methodology for Modelling of Surface Potential, Ionization and Breakdown of Graphene Field-Effect Transistors

Abstract This chapter addresses the methodology used in this thesis, which is divided into three sections. Section 3.1 presents models for surface potential, lateral electric field and length of saturation velocity region (LVSR) of single- and double-gate GNRFETs. Section 3.2 proposes a model for ionization coefficient of GNR, and finally, Sect. 3.3 presents analytical approaches to calculate breakdown voltage of single- and double-gate GNRFETs. In addition, some parts of the results are presented here for the purpose of clarification and will not be repeated in the results and discussion chapter.

Keywords Length of velocity saturation region · Surface potential · Lateral electric field · Ionization · Analytical modelling

3.1 Length of Velocity Saturation Region (L_d)

We divide this section into two subsection. The first section presents the proposed model for length of velocity saturation of a typical single-gate GNRFET. Then, we extend the model to double-gate GNRFET, which is presented in the second subsection.

3.1.1 1D Model for Single-Gate GNRFET

A schematic cross section of top-gated GNRFET is shown in Fig. 3.1, where t_{ox} is the oxide thickness of top gate with dielectric constant of ε_{ox}; t_g, W and L are the GNR's thickness, width and the channel length, respectively.

The channel is divided into two sections. Section 1 is defined between drain and saturation point and Section 2 between saturation point and source junction. We begin with applying Gauss's law in Section 1, shown in Fig. 3.1.

© The Author(s) 2018

I.S. Amiri and M. Ghadiry, *Analytical Modelling of Breakdown Effect in Graphene Nanoribbon Field Effect Transistor*, SpringerBriefs in Applied Sciences and Technology, https://doi.org/10.1007/978-981-10-6550-7_3

Fig. 3.1 Schematic cross section of a top-gated GNRFET

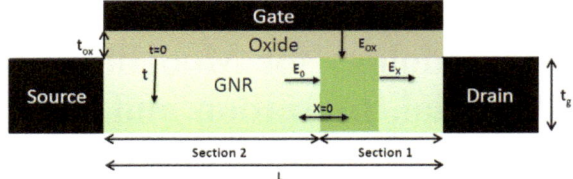

$$-q \int_0^x \int_0^{t_s} (n+N)\mathrm{d}x\mathrm{d}t = -\int_0^x \varepsilon_{\mathrm{ox}}\xi_{\mathrm{ox}}\mathrm{d}x - \int_0^{t_g} \varepsilon_g\xi_0\mathrm{d}t + \int_0^{t_g} \varepsilon_g\xi_x\mathrm{d}t \qquad (3.1)$$

where q is the charge magnitude, ε_g and $\varepsilon_{\mathrm{ox}}$ are the graphene and oxide dielectric constants, n is the intrinsic carrier concentration, N is doping concentration, ξ_{ox}, ξ_0 and ξ_x are the oxide, saturation and lateral electric fields. Taking derivation over 3.1 yields

$$\frac{\partial^2 \phi(x)}{\partial x^2} + \frac{V_g - F_{\mathrm{FB}} - \phi(x)}{\lambda^2} = \frac{q(N+n)}{\varepsilon_g} \qquad (3.2)$$

where $\phi_1(x)$ is the surface potential of GNR at any point along the x-direction inside Section 1 ($0 < x < L_d$), and V_g is the gate voltage. The parameter $\lambda = \sqrt{\varepsilon_g t_g t_{\mathrm{ox}}/\varepsilon_{\mathrm{ox}}}$ is relevant length scale for potential variation [1]. Flat band voltage V_{FB} in GNR with a bandgap $E_g = h\upsilon_F/3w$ is written as Zhang et al. [2]

$$V_{\mathrm{FB}} = \frac{h\upsilon_F}{6qW} - V_T \ln\left(\frac{N}{n}\right) \qquad (3.3)$$

where $\upsilon_F \approx 10^6$ m/s is the Fermi velocity and $V_T = KT/q$ is the thermal voltage. Since quantum capacitance is significant in nanoscale transistors, we need to include the effect of that in the surface potential model. As the enclosed charge in the gauss surface is given by

$$Q = q \int_0^{t_g} \int_0^x (n+N)\mathrm{d}t\mathrm{d}x, \qquad (3.4)$$

$q(n+N)$ can be replaced by $Q/t_g x$. Therefore, surface potential is written as

$$\frac{\partial^2 \phi_1(x)}{\partial x^2} + \frac{V_g - F_{\mathrm{bi}} - \phi_1(x)}{\lambda^2} = \frac{Q}{x t_g \varepsilon_g} \qquad (3.5)$$

The charge Q can also be calculated from

$$Q = (C_g + C_q)(V_{FB} + \phi_{ch} - V_{sub})t_g, \qquad (3.6)$$

where $C_g = \frac{\varepsilon_g}{t_g}$ is the GNR capacitance, C_q is the quantum capacitance of the channel, V_{sub} is the substrate voltage and ϕ_{ch} is surface potential in the central region of the channel. The concentration of the holes has been neglected here. Capacitance C_q is the quantum capacitance of the channel which is given by

$$C_q = q^2 \frac{\partial n}{\partial E}, \qquad (3.7)$$

where E is the energy. The ϕ_{ch} can be approximated to $\phi(L/2)$ where $\phi(x)$ is the surface potential at any point along the channel. This term will be addressed later in this paper. The two-dimensional carrier concentration n_{2D} is written as [3]

$$n_{2D} = \int_{0}^{+\infty} \mathrm{DOS}(f(E - E_{F_d}) - f(E - E_{F_s}))\mathrm{d}E \qquad (3.8)$$

$$f(E - E_{F_i}) = \frac{1}{1 + \exp\left(\frac{E - E_{F_i}}{KT}\right)} \qquad (3.9)$$

where for $i = s, d$.

We approximate $E_{F_s} = E_F$ and $E_{F_d} = E_F - qv_{ds}$ [3, 4], where Fermi energy is written as $E_F = qV_{Vh}$ with V_{ch} being the channel potential Fang et al. [5]. The density of states DOS is given by [6] as

$$\mathrm{DOS} = \frac{2m_e}{\pi\hbar^2} \qquad (3.10)$$

$$\zeta_i = \frac{1}{1 + \exp\left(\frac{E - E_{F_i}}{KT}\right)}$$

Replacing

$$n_{2D} = \int_{0}^{+\infty} \mathrm{DOS}(\zeta_d - \zeta_s)\mathrm{d}E \qquad (3.11)$$

Finally, $n = n_{2D} \frac{1}{t_g + t_{int}}$ with t_{int} being the interlayer distance of graphene is written and limited as

$$n = \left[\frac{1}{t_g + t_{int}}\right] \frac{2m_e}{\pi\hbar^2} \int_0^{\eta_d} \zeta_d dE - \int_0^{\eta_s} \zeta_s dE \tag{3.12}$$

where $\eta_d = \frac{E_{F_i} - E_g/2}{KT}$. Finally, quantum capacitance is given as

$$C_q = q^2 \left[\frac{1}{t_g + t_{int}}\right] \frac{2m_e}{\pi\hbar^2} (\zeta_d - \zeta_s) \tag{3.13}$$

Now, we can proceed to calculate surface potential and lateral electric field analytically. Boundary conditions for $\phi_1(x)$ are defined as $\phi_1(0) = V_0 + V_{bi}$, $\phi_1(L_d) = V_{bi} + V_{ds}$, $\xi_1(0) = \xi_0$, where V_0, V_{ds}, L_d, V_{bi} and ξ_0 are the saturation voltage at the onset of saturation region, drain voltage, length of saturation region, bulk-drain built-in voltage and saturation surface electric field, respectively. Solving the differential equation and taking

$$A = \frac{Q}{t_g x \varepsilon_g} - \frac{V_g - V_{bi}}{\lambda^2} \tag{3.14}$$

yields

$$\phi_1(x) = \lambda^2 A \left(\cosh\left(\frac{x}{\lambda}\right) - 1\right) + (V_0 + V_{bi}) \cosh\left(\frac{x}{\lambda}\right) + \lambda \xi_0 \sinh\left(\frac{x}{\lambda}\right) \tag{3.15}$$

Since $x_1(x) = -\partial\phi_1(x)/\partial x$, surface electric field distribution $\xi_1(x)$ is expressed as

$$\xi_1(x) = -\left(\lambda A + \frac{V_0 + V_{bi}}{\lambda}\right) \sinh\left(\frac{x}{\lambda}\right) - \xi_0 \cosh\left(\frac{x}{\lambda}\right) \tag{3.16}$$

In order to model the surface potential $\phi_2(x)$ between source and saturation points (Section 2), we apply Gauss's law at the region 2 with boundary conditions of $\xi_2(0) = \xi_0$ and $\phi_2(0) = V_0 + V_{bi}$. Assuming that $L_d < L/2$, $\phi(L/2) = \phi_2(L/2 - L_d)$. As a result, ϕ_{ch} is expressed as

$$\phi_{ch} = \cosh\left(\frac{L - 2L_d}{2\lambda}\right) \lambda^2 A - \lambda^2 A + V_0$$
$$+ V_{bi} \cosh\left(\frac{L - 2L_d}{2\lambda}\right) + \lambda \xi_0 \sinh\left(\frac{L - 2L_d}{2\lambda}\right) \tag{3.17}$$

To calculate L_d, Eq. 3.15 can be numerically solved at $x = L_d$. As a result, the effective channel length $L_E = L - L_d$ is given as

$$L_E = L - \frac{\sinh\left(\frac{L_d}{\lambda}\right)V_0\lambda}{V_{ds}\lambda^2 A\left(\cosh\left(\frac{L_d}{\lambda}\right) - 1\right) - (V_0 + V_{bi})\cosh\left(\frac{L_d}{\lambda}\right)} \tag{3.18}$$

According to [1, 7], the electric field at Section 2 can be assumed to be linear. As a result, it is concluded that $\partial^2\phi_2(x)/\partial x^2 = -\xi_0/L_E$. In addition, it is assumed that $\phi_2(0) = V_0 + V_{bi} = \xi_0.L_E + V_{bi}$. Therefore, using Poisson's equation again, ξ_0 is given as

$$\xi_0 = -\frac{L_E\lambda^2}{L_E^2 + \lambda^2}\left(\frac{Q}{\varepsilon_g t_g x} - \frac{V_g - 2V_{bi}}{\lambda^2}\right) \tag{3.19}$$

and V_0 is written as

$$V_0 = -\frac{L_d L_E\lambda^2}{L_E^2 + \lambda^2}\left(\frac{Q}{\varepsilon_g t_g x} - \frac{V_g - 2V_{bi}}{\lambda^2}\right) \tag{3.20}$$

3.1.2 1D Model for Double-Gate GNRFET

Now, we move to double-gate GNRFET surface potential 1D model. At the end, it is possible to conduct a comparison between single- and double-gate GNRFETs in terms of breakdown voltage. A schematic cross section of double-gate GNRFET is shown in Fig. 3.2. Applying Gauss's law on the device presented in Fig. 3.2 yields

$$-\int_0^X \varepsilon_{ox}\xi_{ox1}dx - \int_0^X \varepsilon_{ox}\xi_{ox2}dx - \int_0^{t_g} \varepsilon_g\xi_{ox2}dt + \int_0^{t_g} \varepsilon_g\xi_x dt = q\int_0^x\int_0^{t_g}(n+N)dxdt \tag{3.21}$$

By solving Eq. 3.21 and taking $\xi_{ox} = \frac{V_g - V_{bi} - \phi(x)}{t_{ox}}$, Eq. 3.22 is obtained.

Fig. 3.2 Schematic cross section of a double-gate GNRFET (DG-GNRFET). The parameters ξ_{sat} and ξ_0 are used interchangeably in this text

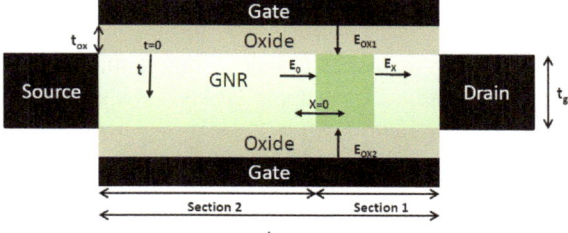

$$\frac{\partial^2 \phi_1(x)}{\partial x^2} + \frac{V_g - V_{bi} - \phi_1(x)}{2\lambda^2} = \frac{q(N+n)}{\varepsilon_g} \tag{3.22}$$

where λ is given as $\lambda = \left(\varepsilon_g t_g t_{ox}/\varepsilon_{ox}\right)^{-2}$.

To include the effect of quantum capacitance in the surface potential model, the same approach as what was used in SG-GNRFET can be applied.

$$Q = q \int\limits_0^x \int\limits_0^{t_g} (n+N)\,dx\,dt \tag{3.23}$$

The term $q(n+N)$ can be replaced by $Q/t_g x$. Therefore, surface potential is written as

$$\frac{\partial^2 \phi_1(x)}{\partial x^2} + \frac{V_g - V_{bi} - \phi_1(x)}{2\lambda^2} = \frac{Q}{\varepsilon_g t_g x} \tag{3.24}$$

Boundary conditions for Eq. 3.22 can be defined as $\phi_1(0) = V_0 + V_{bi}$, $V(L_d) = v_D + V_{bi}$ and $\xi(0) = \xi_S$, where ξ_S, V_S, V_D and L_d are saturation electric field, drain saturation voltage, drain voltage and length of saturation velocity region (LVSR), respectively [7]. Solving Eq. 3.22 with the defined boundary conditions and taking

$$A = \frac{Q}{\varepsilon_g t_g x} - \frac{V_g - V_{bi}}{2\lambda^2} \tag{3.25}$$

yield

$$\phi_1(x) = 2\lambda^2 A\left[\cosh\left(\frac{x}{\sqrt{2}\lambda}\right) - 1\right] + (V_0 + V_{bi})\cosh\left(\frac{x}{\sqrt{2}\lambda}\right) + \sqrt{2}\lambda\xi_0 \sinh\left(\frac{x}{\sqrt{2}\lambda}\right) \tag{3.26}$$

Since $\xi_1(x) = -\partial\phi_1(x)/\partial x$, surface electric field distribution $\xi_1(x)$ can be written using as

$$\xi_1(x) = \left(-\sqrt{2}\lambda A - \frac{V_0 + V_{bi}}{\sqrt{2}\lambda}\right)\sinh\left(\frac{x}{\sqrt{2}\lambda}\right) - \xi_0 \cosh\left(\frac{x}{\sqrt{2}\lambda}\right) \tag{3.27}$$

In order to model the surface potential $\phi_2(x)$ between source and saturation points (Section 2), we apply Gauss's law at the region 2 with boundary conditions of $\xi_2(0) = \xi_0$ and $\phi_2(0) = V_0 + V_{bi}$.

Assuming that $L_d < L/2$, $\phi(L/2) = \phi_2(L/2 - L_d)$. As a result, ϕ_{ch} can be expressed as

$$\phi_{ch} = \cosh\left(\frac{L - 2L_d}{2\sqrt{2}\lambda}\right)2\lambda^2 A - 2\lambda^2 A + (V_0 + V_{bi})\cosh\left(\frac{L - 2L_d}{2\sqrt{2}\lambda}\right)$$
$$+ \sqrt{2}\lambda\xi_0 \sinh\left(\frac{L - 2L_d}{2\sqrt{2}\lambda}\right) \tag{3.28}$$

In order to calculate L_d, Eq. 3.15 can be numerically solved at $x = L_d$. As a result, the effective channel length $L_E = L - L_d$ is given as

$$L_E = L - \frac{\sinh\left(\frac{L_d}{\sqrt{2}\lambda}\right)V_0\sqrt{2}\lambda}{V_{ds} - 2\lambda^2 A\left[\cosh\left(\frac{L_d}{\sqrt{2}\lambda}\right) - 1\right] - (V_0 + V_{bi})\cosh\left(\frac{L_d}{\sqrt{2}\lambda}\right)} \tag{3.29}$$

To calculate V_0, lateral electric field at saturation is modelled first. The electric field at Section 2 can be assumed to be linear according to [1, 7, 8]. As a result, it can be concluded that $\partial^2\phi(L_E)/\partial x^2 = -\xi_0/L_E$. Therefore, using Poisson's equation again, ξ_0 is given as

$$\xi_0 = -\frac{2L_E\lambda^2}{L_E^2 + 2\lambda^2}\left(\frac{Q}{\varepsilon_g t_g x} - \frac{V_g - 2V_{bi}}{2\lambda^2}\right) \tag{3.30}$$

and V_0 is written as

$$V_0 = -L_d\frac{2L_E\lambda^2}{L_E^2 + 2\lambda^2}\left(\frac{Q}{\varepsilon_g t_g x} - \frac{V_g - 2V_{bi}}{2\lambda^2}\right) \tag{3.31}$$

3.1.3   2D Model for Double-Gate GNRFET

In this thesis first, it was tried to model the surface potential using a two-dimensional approach. However, the approach was not further pursued because of two reasons. Firstly, due to complexity of calculations, time-consuming simulation was resulted. Secondly, the aim of this thesis is to model the surface potential and lateral electric field (only at surface) meaning that a simple one-dimensional approach is enough. However, since we have done the simulations, it is worth to mention them and the derived equations in this section.

A schematic cross section of DG-GNRFET is shown in Fig. 3.3, where t_{ox} is the oxide thickness of front and back gates with dielectric constant of ε_{ox}. The t_g, ε_g, W and L are the thickness, dielectric constant, width and length of the GNR, respectively. Generally, to model the potential distribution, Poisson's equation is solved [9].

Fig. 3.3 Schematic cross section of a double-gate GNRFET

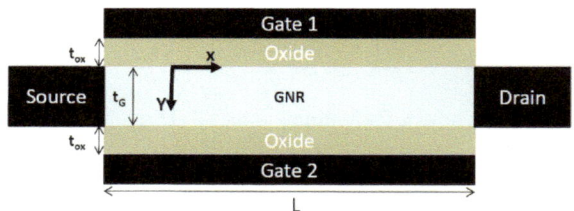

$$\nabla^2 \phi(x, y) = -\frac{qN}{\varepsilon_g}, \quad 0 \le x \le t_g, \ 0 \le y \le L \tag{3.32}$$

where $\phi(x, y)$ is the potential at any point (x, y), in the GNR, q is the electric charge magnitude and N is the doping concentration of GNR. Ignoring the built-in potential of the source/drain channel junction [9], the boundary conditions of Eq. 3.32 are defined as $\phi(0, 0) = V_{bi} + 0$ and $\phi(0, L) = V_{bi} + V_{DS}$. In addition, as the electric flux along the front and back GNR/oxide interface is continuous, the potential function must satisfy

$$\frac{\partial^2 \phi(0, y)}{\partial x} = \frac{\varepsilon_{ox}}{\varepsilon_g} \times \frac{\phi(0, y) - V_{g1}}{t_{ox}} \tag{3.33}$$

And

$$\frac{\partial \phi(t_g, y)}{\partial x} = \frac{\varepsilon_{ox}}{\varepsilon_g} \times \frac{V_{g2} - \phi(t_g, y)}{t_{ox}} \tag{3.34}$$

where $V_{g1} = V_{GS1} - V_{FB1}$, $V_{g2} = V_{GS2} - V_{FB2}$, V_{GS1} and V_{GS2} are gate-source voltages for front and back gates, respectively, and V_{FB1} and V_{FB2} are front and back flat band voltages, respectively. Flat band voltage V_{FB} in GNR with a bandgap $E_g = h\upsilon_F/3w_g$ [2] can be calculated using Eq. 3.35

$$V_F = \frac{h\upsilon_F}{6qw_g} - V_T \ln \frac{N}{n} \tag{3.35}$$

where υ_F 10^6 m/s is the Fermi velocity of graphene, $V_T = K_B T/q$ is the thermal voltage, n is the intrinsic carrier concentration of graphene. According to [9, 10], $\phi(x, y)$ can be decomposed into two parts:

$$\phi(x, y) = V(x) + U(x, y) \tag{3.36}$$

where $V(x)$ is the 1D solution of the Poisson's equation

$$\frac{\partial V(x)}{\partial x^2} = \frac{-qN}{\varepsilon_g} \qquad (3.37)$$

which accounts for long-channel effects and $U(x, y)$ is the solution of Poisson's equation which deals with 2D short-channel effects. Using Eqs. 3.36 and 3.37 in 3.32, $U(x, y)$ satisfies the Laplace equation

$$\frac{\partial^2 U(x, y)}{\partial x^2} + \frac{\partial^2 U(x, y)}{\partial xy^2} = 0 \qquad (3.38)$$

As Eq. 3.36 shows, the boundary conditions of $\phi(x, y)$ can be also written into two parts. Therefore, by separating Eqs. 3.33 and 3.34, the boundary conditions can be expressed as

$$\frac{\partial V(x)}{\partial x} = \frac{\varepsilon_{ox}}{\varepsilon_g} \left(\frac{V(0) - V_{g1}}{t_{ox}} \right) \qquad (3.39)$$

and

$$\frac{\partial V(x)}{\partial x} = \frac{\varepsilon_{ox}}{\varepsilon_g} \left(V_{g2} - \frac{V(0)}{t_{ox}} \right) \qquad (3.40)$$

Solving Eq. 3.37 with boundary conditions of Eqs. 3.39 and 3.40 yields

$$V(0) = V_0 = \frac{C_g}{2C_g + C_{ox}} \left[V_{g2} + V_{g1} \left(\frac{C_{ox}}{C_g} + 1 \right) + qNt_g \left(\frac{1}{C_g} + \frac{1}{C_{ox}} \right) \right] \qquad (3.41)$$

The boundary conditions for Eq. 3.38 are expressed as

$$U(0, 0) = -V_0 \qquad (3.42)$$

$$U(0, L) = V_{DS} - V_0 \qquad (3.43)$$

$$\frac{\partial U(0, y)}{\partial x} = \frac{\varepsilon_{ox}}{\varepsilon_g} \frac{U(0, y)}{t_{ox}} \qquad (3.44)$$

$$\frac{\partial U(t_{ox}, y)}{\partial x} = \frac{\varepsilon_{ox}}{\varepsilon_g} \frac{U(t_{ox}, y)}{t_{ox}} \qquad (3.45)$$

The solution of Eq. 3.38 can be obtained by the separation of variables method [10]. The solution at the surface is given by the following exponential series

$$U(0, y) = \sum_{n=1}^{\infty} A_n \exp(\lambda_n y) + B_n \exp(-\lambda_n y) \tag{3.46}$$

where

$$A_n = \frac{(V_{DS} - V_0) \exp(-\lambda_n L) + V_0 \exp(-\lambda_n L)}{1 - \exp(-\lambda_n L)} \tag{3.47}$$

$$B_n = -V_s - A_n \tag{3.48}$$

and λ_n is potential variations parameter defined as

$$t_g \lambda_n = \left(\frac{C_g}{2C_{ox}}\right)\left[(t_g \lambda_n)^2 - \left(\frac{C_{ox}}{C_g}\right)\right] \tan(t_g \lambda_n) \tag{3.49}$$

As t_g is a small value (in order of 10^{-9}), $\tan(t_g \lambda_n)$ can be approximated to $t_g \lambda_n$. Thus, λ_n is given by

$$\lambda_n = \frac{1}{t_g}\sqrt{1 + \frac{2C_{ox}}{C_g}} \tag{3.50}$$

In order to get a simple solution of $U(0, y)$, it can be approximated to only the first term ($n = 1$) of the series in Eq. 3.46 according to [9]. Thus, Eq. 3.46 reduces to Eq. 3.51 taking into account that we approximated $\exp(-x) \approx 0$ for $x > 3$. This approximation is justified if y and $L > 3\lambda$, which is relevant in this study.

$$U(0, y) = (V_{DS} - V_0) \exp(\lambda(y - L)) - V_0 \exp(-\lambda y) \tag{3.51}$$

Thus, using Eq. 3.36, the surface potential along y can be written as

$$\phi(0, y) = V_0 + (V_{DS} - V_0) \exp(\lambda(y - L)) - V_0 \exp(-\lambda y) \tag{3.52}$$

In addition, the lateral electric field along the channel can be expressed as derivation of Eq. 4.6 over y.

$$\xi(0, y) = \lambda[(V_{ds} - V_0) \exp(\lambda(y - L)) + V_0 \exp(-\lambda y)] \tag{3.53}$$

By taking $y = L - L_d$, $\phi(0, y) = V_{Sat}$, and solving Eq. 4.6 for L_d, we have

$$L_d = L - \frac{1}{\lambda}\ln\left(\frac{\frac{V_{ds} - V_0}{\exp(\lambda L)}\exp(2\lambda(L - L_d)) - V_0}{V_{sat} - V_0}\right) \tag{3.54}$$

which can be solved numerically. In Eq. 4.8, V_{sat} and l_d are drain saturation voltage and length of saturation velocity region, respectively. The proposed equations simply explain the relations of surface potential, electric field and length of saturation region with t_{ox}, t_g, V_{DS} and L.

3.2 Ionization Coefficient

This section provides a model relying on lucky drift theory (LD) for ionization coefficient of GNR. It is organized as follows. Sect. 3.2.1 verifies the assumptions made in this modelling. Section 3.2.2 derives and explains the proposed ionization model and finally Sect. 3.2.3 proposes a semianalytical approach to calculate ionization threshold energy.

3.2.1 Verifying the Models' Assumptions

As mentioned in the previous section, there are several research efforts providing analytical models for calculation of ionization coefficient in silicon. Before starting the modelling, the possibility of applying silicon's ionization models in GNR is studied using analytical modelling and simulation. Three main assumptions have been made in the most important proposed models for ionization in silicon [11–15]. First, it has been assumed that the bandstructure is parabolic. Second, the momentum and energy mean free path, λ_m and λ_E, are considered independent of energy E. Third, the steady-state drift velocity has been assumed to be independent of energy. To continue, the validity of each one of the assumptions is studied in GNR.

GNR has a parabolic bandstructure. According to [16], in general, the larger the bandgap that opens in a GNR, the more the valence and conduction bands become parabolic (rather than cone-shaped). As in this thesis GNR is used as FET channel with a nonzero bandgap, its bandstructure is considered as parabolic.

Momentum mean free path and time are functions of energy in GNR. Momentum mean free path is calculated from $\lambda_m = v_F \tau_m$ [17], where τ_m is the momentum relaxation mean free time and $v_F = 1 \times 10^6$ m/s is the Fermi velocity. In [18], τ_m as a function of energy is written as

$$\tau_m(E) = \frac{4\hbar^3 \rho_m \left(v_{ph} v_F\right)^2}{D^2 K_B T E} \tag{3.55}$$

where $\rho_m = 7.6 \times 10^{-8}$ g/cm^2 is the mass density of graphene, $v_{ph} = 2 \times 10^4$ m/s is the sound velocity in 2D graphene and $D = 16.5$ eV is the acoustic deformation potential. The analytically calculated values of $1/\tau_m$ and $\lambda_m = \tau_m v_F$ are shown in Fig. 3.4a, b, respectively. Apparently, as carrier's energy increases, the scattering

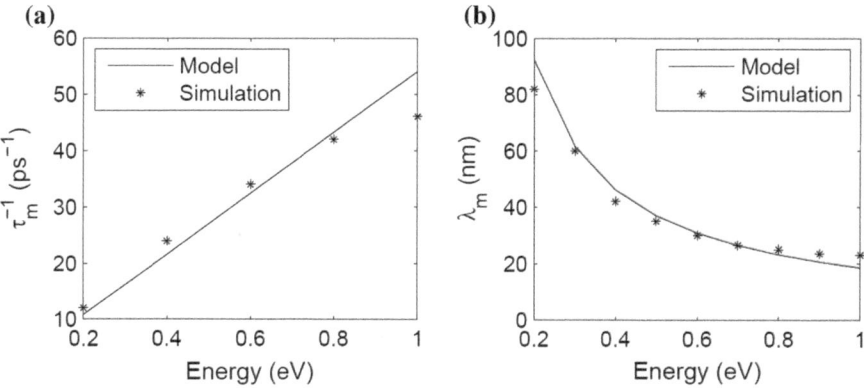

Fig. 3.4 Momentum relaxation mean free time and path (τ_m and λ_m) as a function of energy

rate increases and therefore, the momentum mean free path decreases. The simulation results extracted from [17] also confirm this conclusion.

The values obtained from modelling (Eq. 3.55) and simulation [17] have been compared together in this figure showing that equation $\lambda_m = \tau_m v_F$ is valid for calculation of λ_m.

Drift velocity can be taken as a constant at high energies in GNR. According to [19], at high electric fields, the drift velocity is calculated from

$$v_d = \frac{\mu F}{\left(1 + \left(\frac{\mu F}{v_{\text{sat}}}\right)^\gamma\right)^{\frac{1}{\gamma}}} \tag{3.56}$$

where $\mu = (qn\rho)^{-1}$ is carrier mobility, n is carrier density, ρ is the GNR resistivity [20], $\gamma = 2$ is a fitting parameter and v_{sat} is the saturation velocity given by [19]

$$v_{\text{sat}} = \frac{2\omega_{\text{op}}}{\sqrt{\pi^3 n_2}} \sqrt{1 - \frac{\omega_{\text{op}}^2}{4\pi n_2 v_F^2} \frac{1}{N_{\text{op}} + 1}} \tag{3.57}$$

where $N_{\text{op}} = 1/\exp\left(\frac{\hbar\omega_{\text{op}}}{K_B T}\right)$ is the phonon occupation, ω_{op} is the optical phonon frequency and $n_2 = 5 \times 10^{16}$ m^{-2} is the 2D carrier concentration. To continue, we derive a model for energy to first model the v_{drift} velocity as a function of energy and second to show that electron energy E can go beyond the ionization threshold energy E_t at certain electric field strengths which is the main condition for experiencing ionization.

The average energy of electron is given by $E = qF_g\tau_m$. Since according to [20], the dominant inelastic scattering mechanism in GNR is the phonon scattering, it can be assumed that [11]

$$\tau_{\mathrm{E}} = \frac{E\tau_{\mathrm{m}}(E)}{\hbar\omega_{\mathrm{op}}} \tag{3.58}$$

Using τ_{m} given by Eq. 3.55, average electron energy is written as

$$E = E_0 + E_1 \frac{q\upsilon_{\mathrm{g}}\hbar^2\rho_{\mathrm{m}}\left(\upsilon_{\mathrm{ph}}\upsilon_{\mathrm{F}}\right)^2}{4D^2 K_{\mathrm{B}} T\omega_{\mathrm{op}}} \tag{3.59}$$

where E_0 and E_1 are constant fitting parameters obtained from simulation.

Using Eq. 3.59, energy can be numerically calculated as a function of electric field as shown in Fig. 3.5a, b. In Fig. 3.5a, the energy is plotted with respect to the electric field. It is obvious that as electric field increases, the carrier's energy also increases. It can be also concluded from the proposed model given by Eq. 3.59. We compared the energy values calculated from the proposed model with those extracted from simulation reported in [21], and a good agreement was achieved. In Fig. 3.5b, the profile of energy at high electric field is shown. As can be seen, energy is almost a linear function of electric field. Later, ionization threshold is semianalytically modelled and calculated. The values of ionization threshold voltage will show that ionization is possible in GNR. Using the obtained equation for energy as a function of electric filed, the drift velocity is given as a function of energy as

$$\upsilon_{\mathrm{d}}(E) = \frac{\mu(E - E_0)c_2}{\left(1 + \left(\frac{\mu(E-E_0)c_2}{\upsilon_{\mathrm{sat}}}\right)^\gamma\right)^{\frac{1}{\gamma}}} \tag{3.60}$$

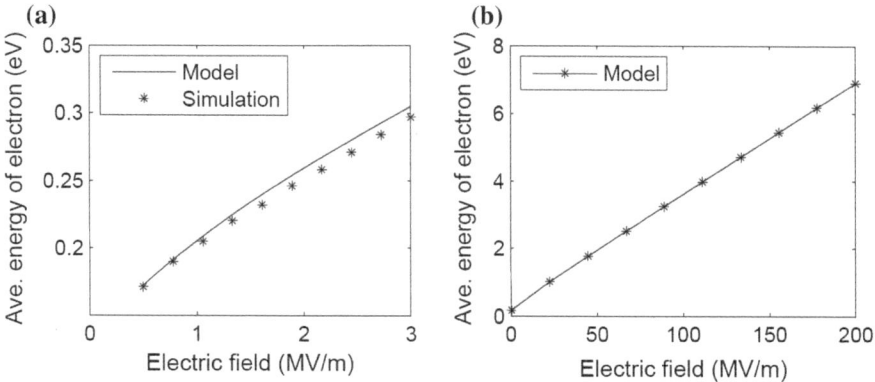

Fig. 3.5 Average energy of electron versus electric field **a** at moderate strength **b** at high strength. Comparison of the modelled results using Eq. 3.59 with simulated data shows that the proposed model agrees well with simulation

Fig. 3.6 Profile of drift
velocity against energy
calculated from Eq. 3.60.
Drift velocity saturates at
3.5×10^5 m/s from 0.4 eV
energy onward

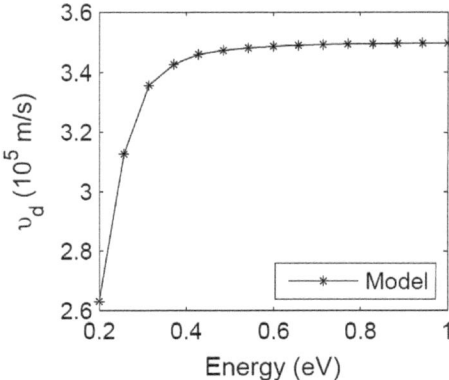

where c_2 is

$$c_2 = E_0 + \frac{q^4 D^2 K_B T \omega_{op}}{E_1 q v_g \hbar^2 \rho_m (v_{ph} v_F)^2} \qquad (3.61)$$

The drift velocity versus energy is plotted in Fig. 3.6. As can be seen, the drift
velocity is strongly related to energy at low energies. However, it reaches steady
state at energies higher than 0.4 eV. As ionization is studied at high energies
(normally more than 0.4 eV), the drift velocity can be assumed independent of the
energy. To sum up, in our model, we assume the bandstructure to be parabolic and
treat τ_m, τ_E, λ_m and λ_E as functions of energy. The drift velocity is assumed to be
independent of energy because ionization coefficient is commonly analysed at high
energies.

3.2.2 The Proposed Model for Ionization Coefficient

It is assumed that carriers can reach threshold energy through ballistic or drift
motions. Figure 3.7 illustrates schematically a typical carrier trajectory and the
associated energy gain [11]. Lateral electric field causes carriers to move semilat-
erally due to collisions deviating them from the lateral dimension.

Collisions could be energy relaxing with rate $1/\tau_E$ due to inelastic collisions or
momentum relaxing with rate $1/\tau_m$ due to elastic collisions. Both the collisions
result in change of direction. Due to each inelastic collision, the carrier loses its
energy as much as optical phonon energy ($\hbar \omega_{op}$). After gaining enough energy (E_t),
ionization could occur which is shown with black star in Fig. 3.7.

We derive the ionization coefficient model in three subsections. The probability
of each one is calculated in subsections 1 and 2, respectively. Then, in the subsection 3, the threshold energy is calculated semianalytically. Using the calculated
probabilities and threshold energy, the ionization coefficient is expressed as

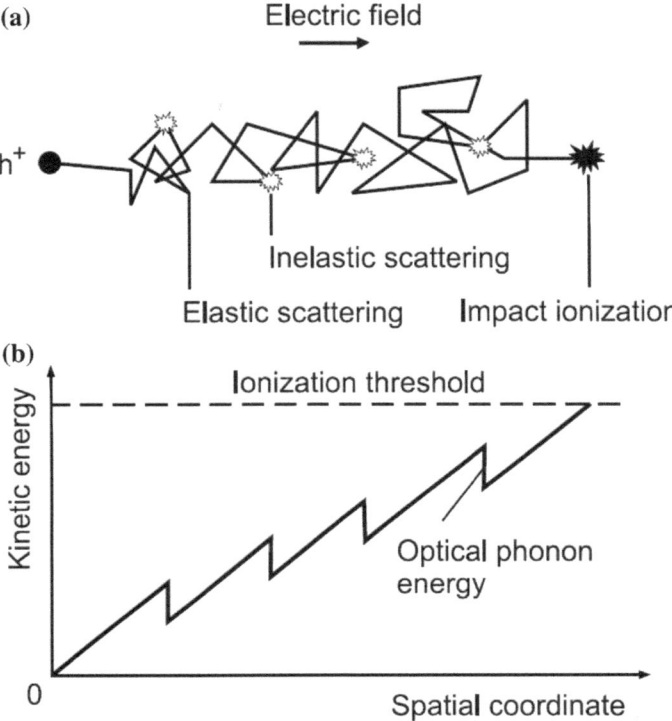

Fig. 3.7 Typical carrier trajectory and the associated energy gain. As carrier experiences the lateral electric field, it gains energy from field. During the path along the direction of electric field, it may lose its energy due to collisions. (Figure extracted from [11])

$$\alpha = \frac{P_{LB}(E,F) + P_{LD}(E,F)}{l_0} \tag{3.62}$$

where $P_{LB}(E,F)$ and $P_{LD}(E,F)$ are the probabilities of reaching threshold energy through ballistic and drift motion, respectively, and $l_0 = E_t/qF$ is the distance travelled by carrier prior to impact ionization assuming no collision is possible.

3.2.2.1 Ballistic Motion

If $P(F,t)$ is the probability that electron avoids significant momentum-relaxing collisions during travel time t, it is true that

$$P_{LB}(F,t) = \exp\left(-\int_0^t \frac{dt}{\tau_m(E)}\right) \tag{3.63}$$

In order to replace t with energy, a relation between them must be found. Electron travelling distance x in an electric field with strength F gains energy equal to $E = qFx$. Taking $x = \lambda_m$ results in energy gain rate $dE/dt = qFv_g$, where according to [20], group velocity $v_g(E)$ is given as

$$v_g = \frac{2}{\pi \hbar \text{DOS}} \tag{3.64}$$

Therefore, $P_{LB}(F, E)$ is given by

$$P_{LB}(F, E_t) = \exp\left(-\int_0^{E_t} \frac{dE}{\tau_m(E) v_g qF}\right) \tag{3.65}$$

Using expression of τ_m in Eq. 3.55, Eq. 3.65 is written as

$$P_{LB}(F, E_t) = \exp\left(-\int_0^{E_t} \frac{4D^2 K_B T E}{\hbar^3 \rho_m \left(v_{ph} v_F\right)^2 v_g qF} dE\right) \tag{3.66}$$

Performing integral, Eq. 3.66 is rewritten as

$$P_{LB}(F, E_t) = \exp\left(\frac{-E_t c_1}{c_1 v_g qF} dE\right) \tag{3.67}$$

where

$$c_1 = \frac{\hbar^3 \rho_m \left(v_{ph} v_F\right)^2}{4D^2 K_B T} \tag{3.68}$$

3.2.2.2 Drift Motion

An electron starting from zero energy may move ballistically up to energy E and then up to E_t having momentum-relaxing collisions. This motion is called lucky drift motion. The probability of not colliding during time t is $P_{LB}(F, t)$, and the probability of colliding during a time dt is $dt/\tau_m(E)$. Thus, the probability of a first collision in the time interval dt after t is just $P_{LB}(F, t)dt/\tau_m(E)$. Converting to energy, it is

$$P_{LD1}(F, E_t) = \frac{P_{LB}(F, E)}{\tau_m(E) v_g qF} dE = P_{LB}(F, E) \frac{E}{c_1 F} \tag{3.69}$$

The probability of lucky drift from E to E_t could be calculated from P_{LD2}.

$$P_{LD2}(F, E) = \exp\left(-\int_E^{E_t} \frac{dE}{\tau_E(E)v_g qF}\right) \tag{3.70}$$

Using the ratio of $\tau_E/\tau_m = E/\hbar\omega_{op}$ according to [11], Eq. 3.70 is rewritten as

$$P_{LD2}(F, E) = \exp\left(-\int_E^{E_t} \frac{E}{c_1 v_g qF} dE\right) \tag{3.71}$$

Finally, the total probability of lucky drift is given by

$$P_{LD}(F, E) = \int_E^{E_t} P_{LD1}(F, E) P_{LD2}(E, F) dE \tag{3.72}$$

Solving the integral analytically gives the $P_{LD}(E, F)$ as

$$P_{LD}(E, F) = c_2\left(P_{LB}(E, F) - \exp\left(\frac{E_1^2}{2c_1 c_1 F}\right) P_{LB}(E, F)\right) \tag{3.73}$$

3.2.3 Ionization Threshold Energy

Now, we need to calculate the ionization threshold energy. According to the reported results for conventional semiconductors, threshold energy is directly related to bandgap energy E_g [22]. In this work, it is assumed that E_t takes a form of $E_t = b(N)E_g$, where $b(N)$ is treated as a variable and N is the number of carbon atoms in traverse direction shown in Fig. 3.8. The bandgap energy in GNR is given by $E_g = 2\pi\hbar v_F/3W(N)$, where $W(N)$ is the GNR's width.

Fig. 3.8 Number of carbon atoms in traverse direction and the vertical distance between two carbon atoms

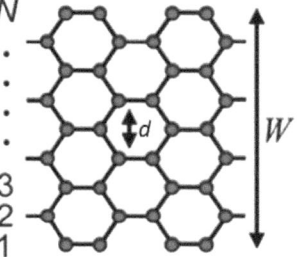

By calculating the distance d shown in Fig. 3.8, GNR's width is written as

$$W(N) = (N - 1)\sqrt{a_{c-c}^2 - (a_{c-c}/2)^2} \tag{3.74}$$

where $a_{c-c} = 0.14$ nm is the carbon–carbon distance. As a result, the threshold energy can be semianalytically computed from

$$E_t = b(N)\frac{2\pi\hbar\upsilon_F}{3(N-1)\sqrt{a_{c-c}^2 - (a_{c-c}/2)^2}} \tag{3.75}$$

3.3 Modelling of Breakdown Voltage

In this section, breakdown voltages for single- and double-gate GNRFETs are modelled using the presented models for the length of saturation region and avalanche integral, and the avalanche condition could be given as

$$1 = \int_0^{L_d} \alpha\big(\xi(V_{ds}, x), W_g, E_t\big)dx \tag{3.76}$$

where the value of V_{ds}, which satisfies the equation, is called BV.

3.3.1 Single-Gate GNRFET

Using the equation derived for surface potential for single-gated graphene nanoribbon FET, the length of saturation region (drift region) was shown to be calculated from

$$L_d = \frac{\sinh\left(\frac{L_d}{\lambda}\right)V_0\lambda}{V_{ds} - \lambda^2 A\left[\cosh\left(\frac{L_d}{\lambda}\right) - 1\right] - (V_0 + V_{bi})\cosh\left(\frac{L_d}{\lambda}\right)} \tag{3.77}$$

Taking $x = L_d$ in Eq. 3.76 and $V_{ds} = BV$ in Eq. 3.77, breakdown voltage BV can be numerically calculated.

3.3.2 Double-Gate GNRFET

The same approach is applied to calculate breakdown voltage in DG-GNRFET numerically using the following equation to calculate L_d.

$$L_d = \frac{\sinh\left(\frac{L_d}{\sqrt{2}\lambda}\right) V_0 \sqrt{2}\lambda}{BV - 2\lambda^2 A \left[\cosh\left(\frac{L_d}{\sqrt{2}\lambda}\right) - 1\right] - V_0 \cosh\left(\frac{L_d}{\sqrt{2}\lambda}\right)} \tag{3.78}$$

References

1. H. Wong, Drain breakdown in submicron MOSFETs: a review. Microelectron. Reliab. **40**(1), 3–15 (2000)
2. Q. Zhang, T. Fang, H. Xing, A. Seabaugh, D. Jena, Graphene nanoribbon tunnel transistors. Electron. Device Lett. IEEE **29**(12), 1344–1346 (2008)
3. M. Cheli, P. Michetti, G. Iannaccone, Model and performance evaluation of field-effect transistors based on epitaxial graphene on SiC. Electron. Devices IEEE Trans. **57**(8), 1936–1941 (2010)
4. G. Liang, N. Neophytou, M.S. Lundstrom, D.E. Nikonov, Computational study of double-gate graphene nano-ribbon transistors. J. Comput. Electron. **7**(3), 394–397 (2008)
5. T. Fang, A. Konar, H. Xing, D. Jena, Carrier statistics and quantum capacitance of graphene sheets and ribbons. Appl. Phys. Lett. **91**(9), 092109 (2007)
6. K.S. Novoselov, A.K. Geim, S. Morozov, D. Jiang, Y. Zhang, S.A. Dubonos, I. Grigorieva, A. Firsov, Electric field effect in atomically thin carbon films. Science **306**(5696), 666–669 (2004)
7. M. El Banna, M. El Nokali, A pseudo-two-dimensional analysis of short channel MOSFETs. Solid-State Electron. **31**(2), 269–274 (1988)
8. D. Krizaj, G. Charitat, S. Amon, A new analytical model for determination of breakdown voltage of Resurf structures. Solid-State Electron. **39**(9), 1353–1358 (1996)
9. M.A. Imam, M.A. Osman, A.A. Osman, Threshold voltage model for deep-submicron fully depleted SOI MOSFETs with back gate substrate induced surface potential effects. Microelectron. Reliab. **39**(4), 487–495 (1999)
10. P.G. Harper, D.L. *Weaire, Introduction to Physical Mathematics: CUP Archive* (1985)
11. O. Rubel, A. Potvin, D. Laughton, Generalized lucky-drift model for impact ionization in semiconductors with disorder. J. Phys. Condens. Matter **23**(5), 055802 (2011)
12. K. Yeom, J. Hinckley, J. Singh, Calculation of electron and hole impact ionization coefficients in SiGe alloys. J. Appl. Phys. **80**(12), 6773–6782 (1996)
13. S. McKenzie, M. Burt, A test of the lucky-drift theory of the impact ionisation coefficient using Monte Carlo simulation. J. Phys. C Solid State Phys. **19**(12), 1959 (1986)
14. B. Ridley, Lucky-drift mechanism for impact ionisation in semiconductors. J. Phys. C Solid State Phys. **16**(17), 3373 (1983)
15. J. Devreese, R. van Welzenis, R. Evrard, Impact ionisation probability in InSb. Appl. Phys. A **29**(3), 125–132 (1982)
16. F. Schwierz, Graphene transistors. Nat. Nanotechnol. **5**(7), 487–496 (2010)
17. W.-K. Tse, E. Hwang, S.D. Sarma, Ballistic hot electron transport in graphene. Appl. Phys. Lett. **93**(2), 023128 (2008)
18. D. Berdebes, T. Low, M. Lundstrom, B.N. Center, *Low Bias Transport in Graphene: An Introduction* (2009)

19. V.E. Dorgan, M.-H. Bae, E. Pop, Mobility and saturation velocity in graphene on SiO_2. arXiv preprint: arXiv:1005.2711 (2010)
20. T. Fang, A. Konar, H. Xing, D. Jena, Mobility in semiconducting graphene nanoribbons: phonon, impurity, and edge roughness scattering. Phys. Rev. B **78**(20), 205403 (2008)
21. R. Shishir, D. Ferry, Velocity saturation in intrinsic graphene. J. Phys. Condens. Matter **21** (34), 344201 (2009)
22. W. Maes, K. De Meyer, R. Van Overstraeten, Impact ionization in silicon: a review and update. Solid-State Electron. **33**(6), 705–718 (1990)

Chapter 4
Results and Discussion on Ionization and Breakdown of Graphene Field-Effect Transistor

Abstract Based on the proposed semi-analytical models in the previous chapter, lateral electric field and length of velocity saturation region are plotted with respect to structural parameters in this chapter. In addition, ionization coefficient is calculated with respect to inverse electric field. Finally, the breakdown voltage is calculated for DG- and SG-GNRFETs and the trends and profiles are discussed. Table 4.1 shows default values for all the parameters which can be used for repeating the experiments.

Keywords Surface potential · Lateral electric field · Saturation region length · Ionization · Analytical results

4.1 Lateral Electric Field and Length of Velocity Saturation Region

In this section, the profile of lateral electric field and length of saturation region of single- and double-gate GNRFETs are plotted with respect to structural parameters such as oxide thickness, channel length and width. In addition, a comparison study is presented to show the effect of the double-gate GNRFET compared to single-gate GNRFET.

4.1.1 Single-Gate GNRFET

As it was seen before, the lateral electric field in single-gate GNRFET is calculated using the following equation

$$\xi(x) = -\left(\lambda\left(\frac{Q}{t_g x \varepsilon_g} - \frac{V_g - V_{bi}}{\lambda^2}\right) + \frac{V_0 + V_{bi}}{\lambda}\right)\sinh\left(\frac{x}{\lambda}\right) - \xi_0\cosh\left(\frac{x}{\lambda}\right) \qquad (4.1)$$

© The Author(s) 2018
I.S. Amiri and M. Ghadiry, *Analytical Modelling of Breakdown Effect in Graphene Nanoribbon Field Effect Transistor*, SpringerBriefs in Applied Sciences and Technology, https://doi.org/10.1007/978-981-10-6550-7_4

Table 4.1 Default value for parameters used in simulation

Parameter name	Value
Carrier concentration	$n_2 = 5 \times 10^{116} \frac{1}{m^2}$
Fermi velocity	$v_F = 10^6$ m/s
Sound velocity	$v_{ph} = 2 \times 10^4$ m/s
Carbon–carbon distance	$a = 0.14 \times 10^{-9}$ m
Charge magnitude	$q = 1.6 \times 10^{-19}$ C
Temperature	300 K
Boltzmann's constant	$K_b = 1.38 \times 10^{-23}$ J/K
Planck's constant	$h = 6.6 \times 10^{-23}$ J s
Reduced Planck's constant	$h/2\pi$ J s
Free electron effective mass	$m^* = 9.11 \times 10$ kg
Electron effective mass in graphene	$m_c = 0.06 m^*$ kg
Hole effective mass in graphene	$m_v = 0.03 m^*$ kg
Phonon energy	$\hbar\omega = 0.2$ eV
Acoustic phonon deformation potential of GNR	$D_{ac} = 16$ eV
Thermal voltage	$V_T = K_B T/q$ V
GNR width	$W = 5$ nm
GNR thickness	$t_g = 0.5$ nm
GNR dielectric constant	$\varepsilon_g = 3.5 \times e_0$ F/m
SiO$_2$ dielectric constant	$\varepsilon_{ox} = 3.9 \times e_0$ F/m
Vacuum permittivity	$\varepsilon_0 = 8.85 \times 10^{-12}$ F/m
Oxide thickness	$t_{ox} = 5$ nm
Channel length	$L = 15$ nm
Gate voltage	$V_g = 0.1$ V
Fitting parameter	$E_0 = 0.31$ eV
Fitting parameter	$E_1 = 0.75$
Fitting parameter	$\gamma = 2$
Optical phonon frequency	$\omega_{op} = 0.2$ eV/$\hbar$

Some values such as channel length and GNR's width may treat as variable

and the length of velocity saturation region is given as

$$L_d = \frac{\sinh\left(\frac{L_d}{\lambda}\right) V_0 \lambda}{V_{ds} - \lambda^2 A\left[\cosh\left(\frac{L_d}{\lambda}\right) - 1\right] - (V_0 + V_{bi})\cosh\left(\frac{L_d}{\lambda}\right)} \qquad (4.2)$$

Figure 4.1 shows the electric field distribution in the lateral direction at different drain-source voltages (V_{ds}). It is worth to mention that we are willing to study the device in saturation condition; thus, $V_{ds} > V_0$, where V_0 is saturation voltage defined as $V_0 = L_d \xi_0$ and ξ_0 is written as

Fig. 4.1 Lateral electric field of SG-GNRFET at different drain voltages. The highest electric field is seen at drain junction

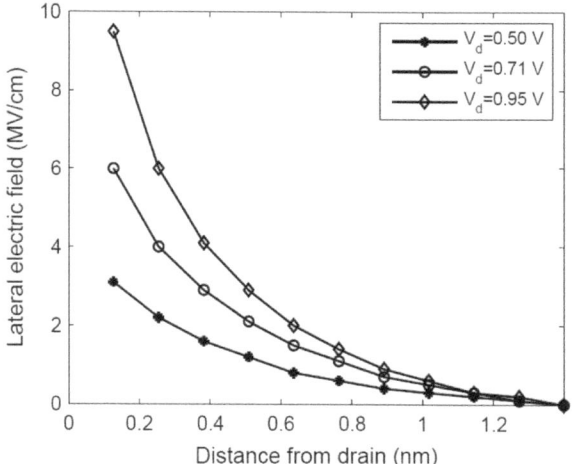

$$\xi_0 = \frac{L_E \lambda^2}{L_E^2 + \lambda^2} \left(\frac{Q}{\varepsilon_g t_g x} - \frac{V_g - 2V_{bi}}{\lambda^2} \right) \tag{4.3}$$

As a result, the values for V_{ds} in Figs. 4.1 and 5.4 are defined as ratio of saturation voltage V_0. Figure 4.1 shows the electric field profile and follows an exponential form depending on the distance from the source.

Based on the electric field model provided, the effect of the drain-source voltage and channel length on the length of saturation region is depicted in Fig. 4.2. The higher the drain voltage and longer channel are, the longer the L_d is. As V_{ds} increases, lateral electric field increases as well. Therefore, carriers travelling from source to drain reach saturation velocity in a shorter distance from source and the length of velocity saturation region increases.

Fig. 4.2 L_d of SG-GNRFET at different L and V_{ds}

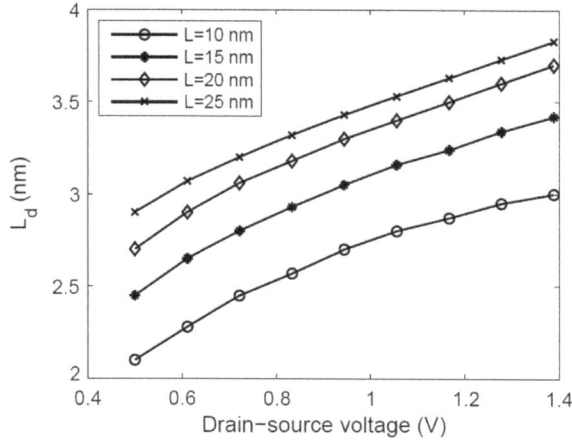

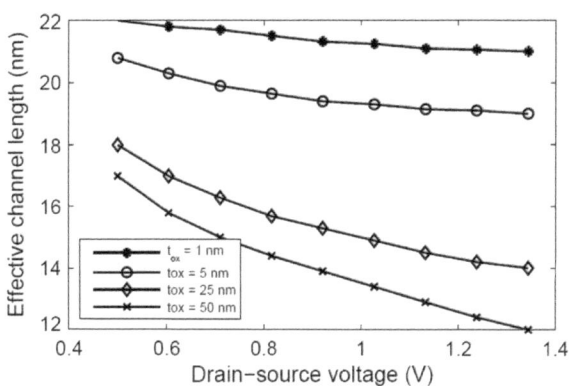

Fig. 4.3 L_d of SG-GNRFET versus t_{ox} variations at different drain voltages

Figure 4.3 shows the oxide thickness effect on the L_d or in the other way, the effective channel, which is $L_E = L - L_d$. It is seen that as thicker oxide is applied, L_E decreases and L_d therefore increases. Increasing oxide thickness results in oxide electric field lowering. Thus, the effective electric field vector's angle gets closer to zero and the carriers are more under the influence of lateral electric field and reach saturation velocity in shorter distance.

4.1.2   Double-Gate GNRFET

4.1.2.1   1D Model

We showed that the lateral electric field in double-gate GNRFET is given as

$$\xi(x) = \left(-\sqrt{2}\lambda\left(\frac{Q}{\varepsilon_g t_g x} - \frac{V_g - V_{bi}}{2\lambda^2}\right) - \frac{V_0 + V_{bi}}{\sqrt{2}\lambda}\right)\sinh\left(\frac{x}{\sqrt{2}\lambda}\right) - \xi_0\cosh\left(\frac{x}{\sqrt{2}\lambda}\right)$$

(4.4)

and the length of saturation region is expressed as

$$L_d = \frac{\sinh\left(\frac{L_d}{\sqrt{2}\lambda}\right)V_0\sqrt{2}\lambda}{BV - 2\lambda^2 A\left[\cosh\left(\frac{L_d}{\sqrt{2}\lambda}\right) - 1\right] - V_0\cosh\left(\frac{L_d}{\sqrt{2}\lambda}\right)}$$

(4.5)

Similar to single-gate GNRFET, the profile of lateral electric field is shown in Fig. 4.4. As it is depicted later, higher lateral electric field is seen in DG-GNRFET compared to that of SG-GNRFET. However, the profile is still the same exponentially reaching peak at drain junction.

Figure 4.5 depicts the effect of the drain voltage and channel length on the length of saturation region. The higher the drain voltage and longer channel are, the

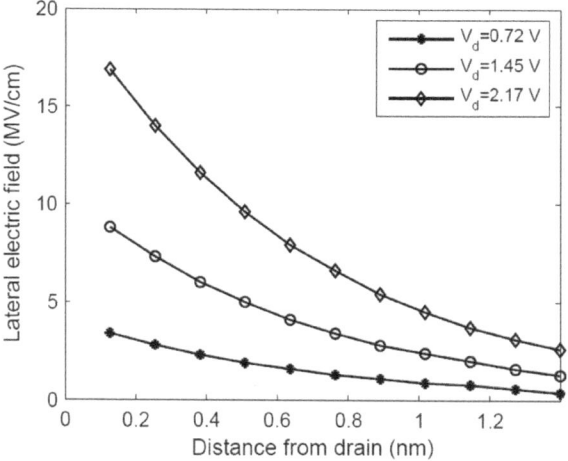

Fig. 4.4 Lateral electric field of DG-GNRFET at different positions from drain junction

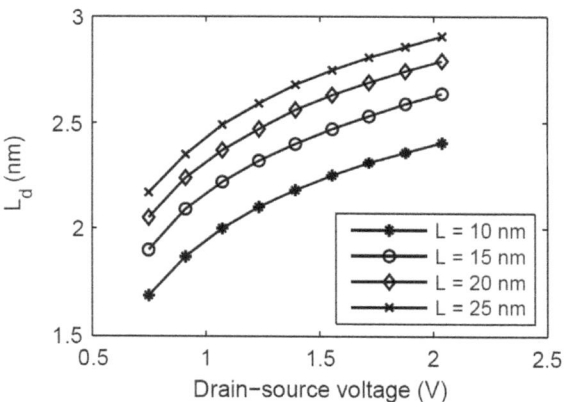

Fig. 4.5 L_d of DG-GNRFET at different channel lengths and drain voltages

longer the L_d is. In Fig. 4.6, it can be seen that increasing the oxide thickness causes increase in L_d. In addition, the term $d(L_d)/d(V_{ds})$ increases as t_{ox} increases.

4.1.2.2 2D Model

In this section, the profile of surface electric field and length of velocity saturation region are shown using the 2D approach presented in Methodology section. In addition, the effect of several parameters such as drain-source voltage, oxide thickness and channel length on the length of saturation region is studied. The surface potential, lateral electric field and length of velocity saturation region are, respectively, given as

Fig. 4.6 L_d of DG-GNRFET versus oxide thickness variations at different drain voltages

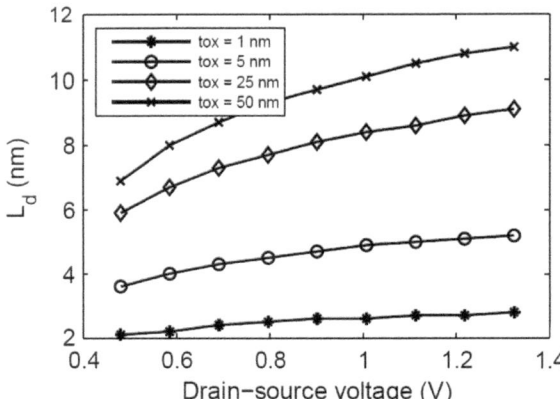

$$\psi(0, y) = (V_{ds} - V_0)\exp(\lambda(y - L)) - V_0\exp(-\lambda y) \tag{4.6}$$

and

$$\xi(0, y) = -\lambda[(V_{ds} - V_0)\exp(\lambda(y - L)) + V_0\exp(-\lambda y)] \tag{4.7}$$

and

$$L_d = L - \frac{1}{\lambda}\ln\left(\frac{\frac{V_{ds} - V_0}{\exp(\lambda y)}\exp(2\lambda(L - L_d)) - V_0}{V_{sat} - V_0}\right) \tag{4.8}$$

For the purpose of model verification, we compared the calculated values using the proposed model with the simulated results by MEDICI for a Si-based device. As shown in Fig. 4.7, good agreement can be seen between the simulation results and model at different doping concentrations, oxide thicknesses and distances from

Fig. 4.7 Comparison of the results extracted from 2D numerical simulator and model

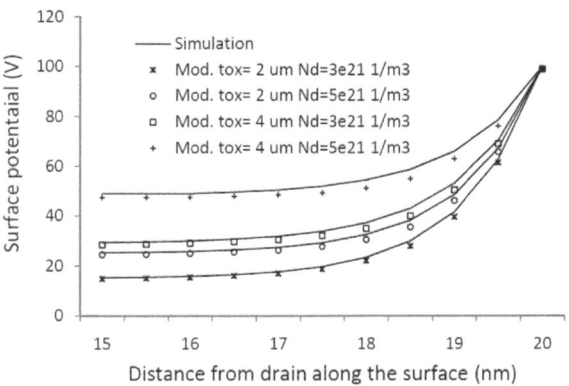

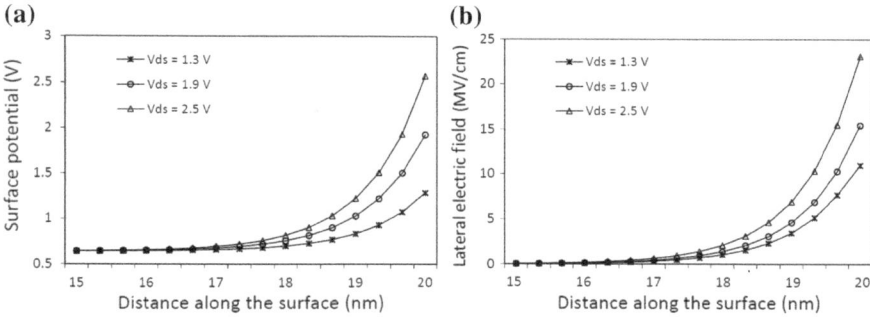

Fig. 4.8 Effect of drain-source voltage on the profile of surface potential and lateral electric field distribution using the proposed model. Default parameters are $N_d = 5 \times 10^{25}$ m^{-3}, $L = 20$ nm, $t_{ox} = 1$ nm and graphene film thickness $t_G = 0.4$ nm

drain. Once the surface potential model is verified, the LVSR model is proved too because it is the solution of the surface potential for $\psi(0, y) = V_{sat}$.

Figure 4.8a, b shows the potential and electric field distribution along the nanoribbon surface is similar to the profile of surface potential in an abrupt junction [1]. Figure 4.9a indicates that increasing the doping concentration in the channel region results in a significant increase in the electric field near the drain junction. Doping concentration has been set to be in order of 10^{25} m^{-3} to be an influential factor in the electric field and length of saturation region. Figure 4.9b shows expanding the oxide thickness causes decrease in LVSR because by doing so, higher saturation voltage and lateral electric filed is resulted, and thus LVSR is shortened. It is worth mentioning that the calculation of LVSR is done for $V_{DS} > V_{sat}$. Therefore, wherever $V_{DS} < V_{sat}$, there is a missing point in the charts. For example, in Figs. 4.9b and 4.10b, LVSR cannot be calculated for a few V_{DSes}. Figure 4.10a shows the dependence of LD on L. As can be seen in this figure and

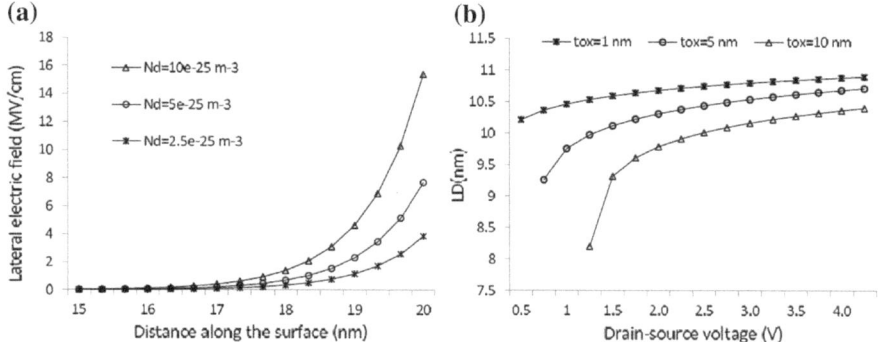

Fig. 4.9 Effect of doping concentration (**a**) and oxide thickness (**b**) on the lateral electric field and length of saturation velocity region. Default parameters are $N_d = 5 \times 10^{25}$ m^{-3}, $L = 20$ nm, $t_{ox} = 1$ nm and $t_G = 0.4$ nm

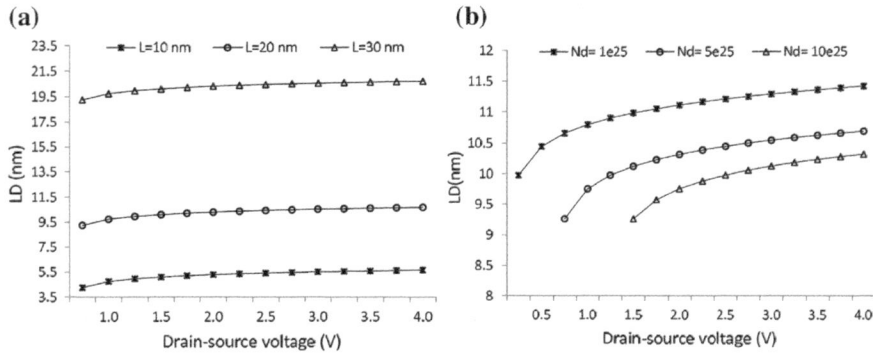

Fig. 4.10 Effect of channel length and doping concentration on the length of saturation velocity region. Default parameters are $N_d = 5 \times 10^{25}$ m^{-3}, $L = 20$ nm, $t_{ox} = 1$ nm and $t_G = 0.4$ nm

Eq. 4.8, there is a direct relation between L and LD. Finally, Fig. 4.10b shows by applying higher doping concentration the saturation voltage increases and LD decreases.

4.1.3 Comparison Between Single- and Double-Gate GNRFETs

In this section, a comparison is made between SG- and DG-GNRFETs in terms of surface potential, lateral electric field and length of saturation region. Figure 4.11 shows that higher lateral electric field is seen in the channel of DG-GNRFET compared to that of SG-GNRFET. Figure 4.12a–c shows length of saturation velocity region of SG- and DG-GNRFETs at different drain-source voltages, oxide

Fig. 4.11 Profile lateral electric field against distance from drain for DG- and SG-GNRFETs. Horizontal axis is the distance from drain in nm

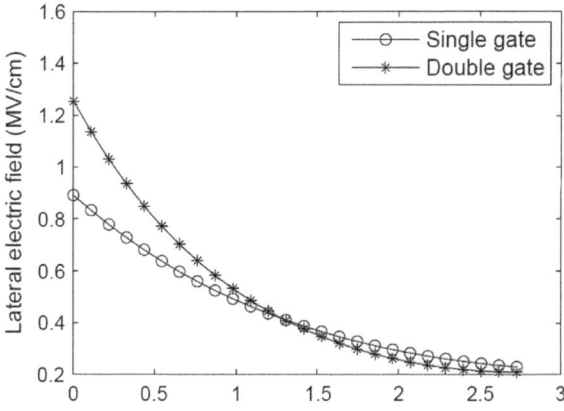

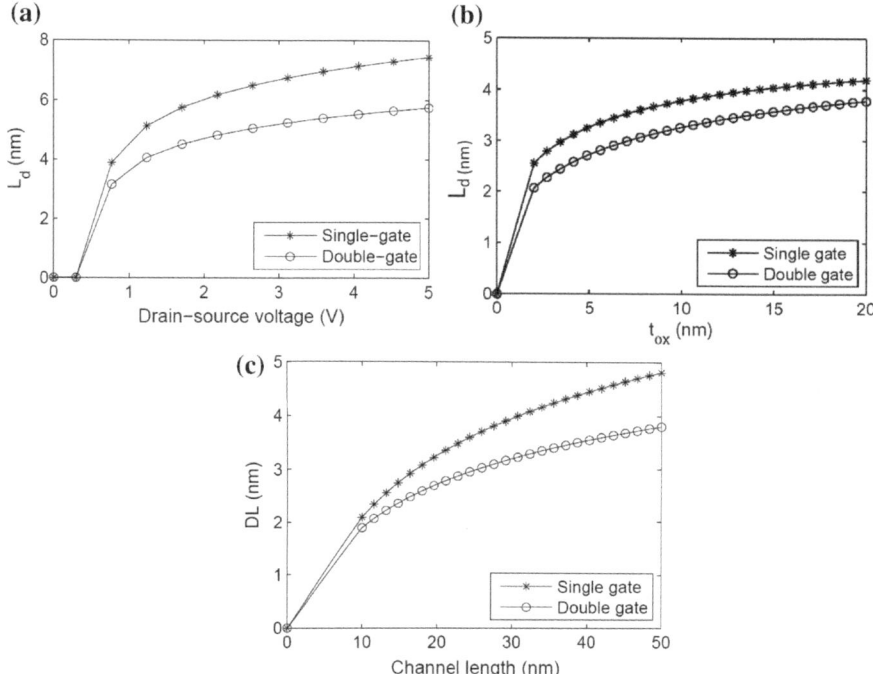

Fig. 4.12 Profile of velocity saturation region length with respect to, drain-source voltage (**a**), oxide thickness (**b**) and channel length (**c**). As can be seen, double-gate FET suppresses the length of drain region

thickness and channel length, respectively. The profiles are almost same and have been explained before.

Comparing SG- and DG-GNRFETs in terms of L_d reveals that the length of drain region is smaller in a double-gate device, because in a double-gate device carriers are under the influence of two vertical electric fields formed by top and back gates; therefore, they need longer path to reach saturation velocity.

In conclusion, using empirical equation extracted from presented charts, the effective parameters on the length of velocity saturation region can simply be given as

$$L_d \propto \frac{L \cdot V_{ds} \cdot t_{ox}}{n_{gate}} \qquad (4.9)$$

where n_{gate} is the number of gates used in the FET, which could be 1 or 2 in SG- and DG-GNRFETs.

4.2 Ionization Coefficient

In this section, ionization coefficient with respect to threshold energy and electric field is plotted. As it was shown in Methodology chapter, ionization coefficient of GNR (α_{GNR}) is given as

$$\alpha_{\mathrm{GNR}} = \frac{P_{LB}(E,F) + P_{LD}(E,F)}{l_0} \tag{4.10}$$

Inserting the equations of $P_{LB}(E, F)$ and $P_{LD}(E, F)$, ionization coefficient is written as

$$\alpha_{\mathrm{GNR}} = \frac{\exp\left(\frac{-E_t^2 c_1}{v_g qF}\right) + \exp\left(\frac{-E_t^2 c_1^2 \hbar\omega_{\mathrm{op}}\left(\hbar\omega_{\mathrm{op}} - E_t\right)}{c_1 v_g qF}\right) \frac{E_t^2}{2c_1} q}{E} \tag{4.11}$$

where the ionization threshold energy was then calculated as

$$E_t = b(N)\frac{2\pi\hbar v_F}{3(N-1)\sqrt{a_{c-c}^2 - (a_{c-c}/2)^2}} \tag{4.12}$$

Based on the proposed equations, Fig. 4.13a, b illustrates the ionization coefficient of GNR at different threshold energies and inverse electric field compared to that of silicon.

In order to verify the results extracted from ionization modelling, we compared the results with results obtained from Monte Carlo simulation reported in recently

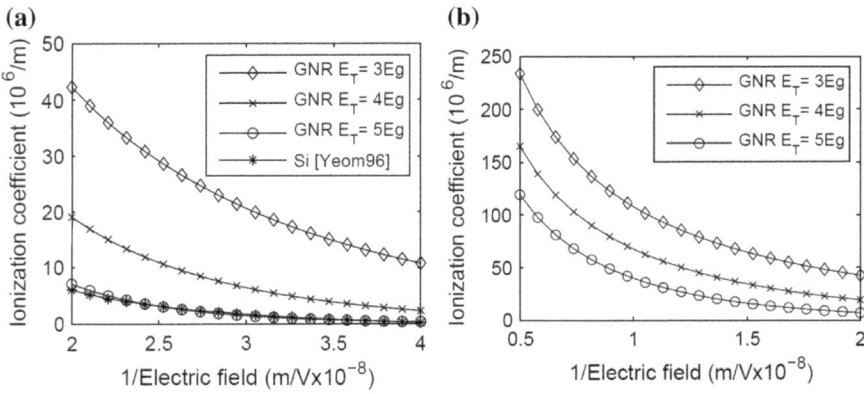

Fig. 4.13 Ionization coefficient of GNR versus reciprocal electric field (**a**) at high strength compared to ionization coefficient of silicon extracted from [2] (**b**) at very high strength

Fig. 4.14 Ionization
coefficient of GNR versus
electric field compared to that
of silicon. Solid lines show
the results of Monte Carlo
simulation from [3]

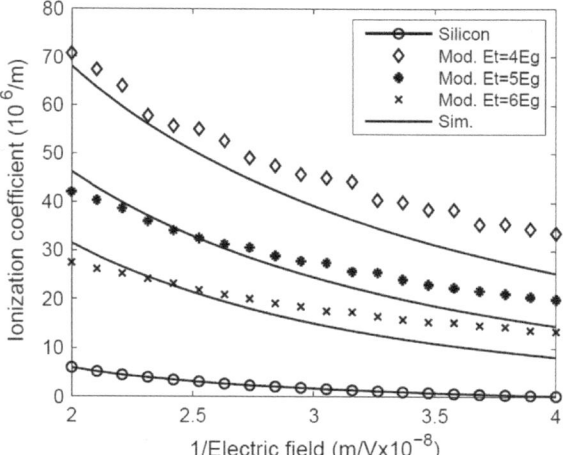

Fig. 4.15 Results of
simulation of threshold
energy using Gaussian atomic
simulator [5] and the ratio of
E_t over E_g

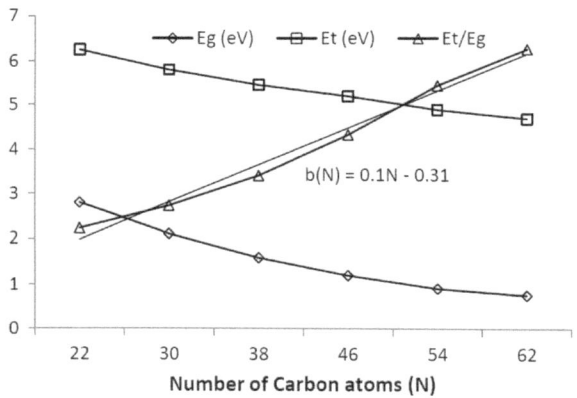

published paper [3]. The results of comparison shown in Fig. 4.14 reveal that the
model agrees well with the simulation results (Fig. 4.15).

In addition, we used Gaussian simulation results for ionization threshold energy
published in [4] to get the ratio of E_t/E_g.

Apparently, the profile of ionization in GNR and silicon is almost same and
ionization coefficient in GNR is comparable to that of silicon. As can be seen,
different threshold energies result in various ionization coefficients. Clearly, as
threshold energy decreases the ionization coefficient increases since it means that
carriers need less energy to jump from valence to conduction band. GNR benefits
from long mean free path (50–400 nm), which means that carriers experience less
collisions than silicon's carriers do. Therefore, they can reach threshold energy
under the influence of lateral electric field faster than silicon carriers do. Thus, the
ionization coefficient of GNR is estimated—using analytical modelling—to be
higher than that of silicon. However, the bandgap of GNR is a tunable parameter;

thus, it is possible to gain less ionization coefficient than silicon for certain applications. In conclusion, ionization coefficient is related to lateral electric field, channel width, bandgap and ionization threshold energy as

$$\alpha \propto \frac{F(x)}{E_g \cdot W \cdot E_t} \tag{4.13}$$

4.3   Modelling Results for Breakdown Voltage

4.3.1   Single-Gate GNRFET

The effect of oxide thickness on breakdown voltage is presented in Fig. 4.16a, b. Apparently, as oxide thickness increases the breakdown voltage increases too. We repeated the simulation for two devices with different channel length to study the role of channel length on the breakdown voltage. As can be seen, increasing the channel length does not affect the breakdown voltage significantly. In these experiments, GNR's width has been taken 5 nm to open a reasonable bandgap in GNR. Figure 4.17a, b shows the breakdown voltage at different GNR's width and channel length. As channel width decreases, the bandgap increases. As a result, carriers need more energy to cross the gap between valence and conduction band and create electron–hole pairs. Thus, ionization threshold energy increases and ionization coefficient decreases.

Therefore, breakdown which is a direct result of ionization happens at higher drain-source voltages and the value of breakdown voltage increases as well. In addition, the figures show that channel length does not have profound influence on the breakdown voltage. This phenomenon can be explained by considering the effect of channel length on L_d and lateral electric field. As channel length increases L_d increases which causes breakdown voltage to increase. However, longer channel

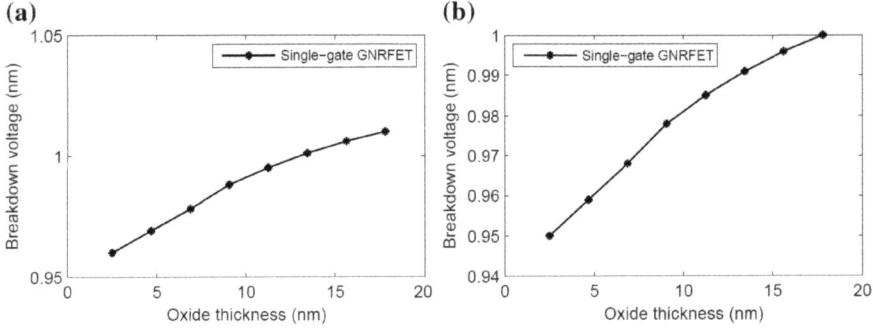

Fig. 4.16 Breakdown voltage SG-GNRFET at different oxide thickness and channel length. **a** $L = 15$ nm, **b** $L = 30$ nm

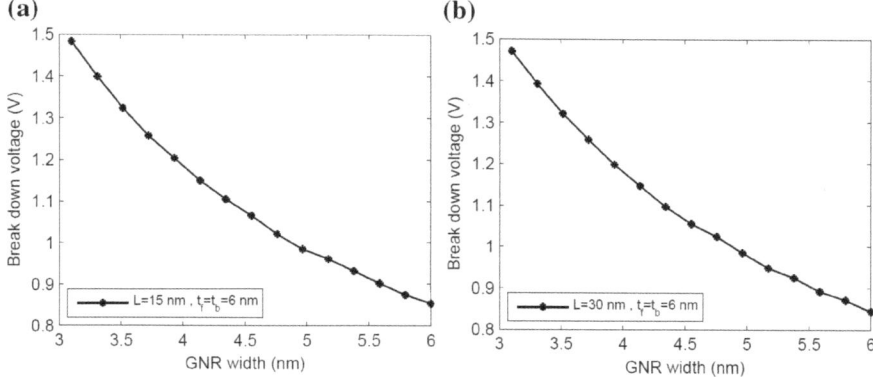

Fig. 4.17 Breakdown voltage for single-gate GNRFET at different GNR's width and channel length

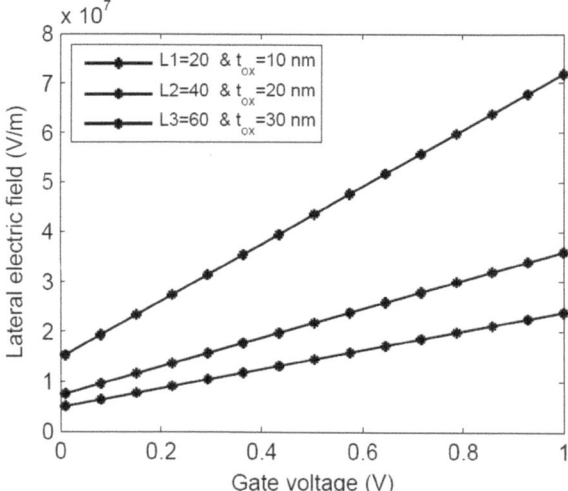

Fig. 4.18 Gate-source voltage versus lateral electric field at the drain region. Different channel length is applied as well as gate-source voltage

length results in less lateral electric field at a fixed V_{ds} meaning less ionization coefficient and thus less breakdown voltage. Therefore, as can be seen in figures, the breakdown voltage does not change significantly by altering channel length.

Finally, the dependence of gate voltage (V_g) is studied on the breakdown voltage in Fig. 4.19. As the gate voltage increases, the breakdown voltage is reduced. The lateral electric field increases by the gate voltage as we show in Fig. 4.18. As a result of higher field, breakdown condition is satisfied at lower drain-source voltage and therefore breakdown voltage decreases.

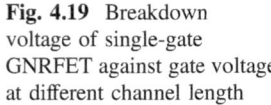

Fig. 4.19 Breakdown
voltage of single-gate
GNRFET against gate voltage
at different channel length

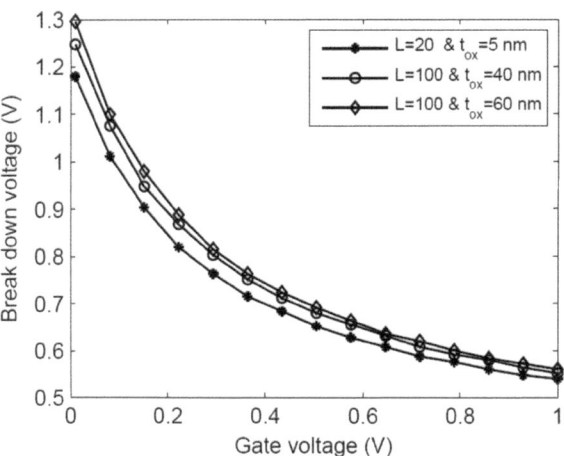

4.3.2 Double-Gate GNRFET

The breakdown voltage of double-gate graphene nanoribbon is calculated using the proposed models, and the effect of oxide thickness, channel length and GNR's width on that is examined. Default conditions of simulation are as follows, $t_f = t_b = 6$ nm, $L = 15$ nm.

Figure 4.20a, b demonstrates the breakdown voltage at various oxide thickness. We assumed the device to be symmetric meaning that $t_f = t_b$. Again, it is seen that increasing the oxide thickness causes breakdown voltage to increase. It reaches a saturation state at almost $t_f = 20$ nm.

Figure 4.21a, b shows the breakdown voltage at different GNR's width and channel length. The result shows that to gain high breakdown voltage, GNR's width

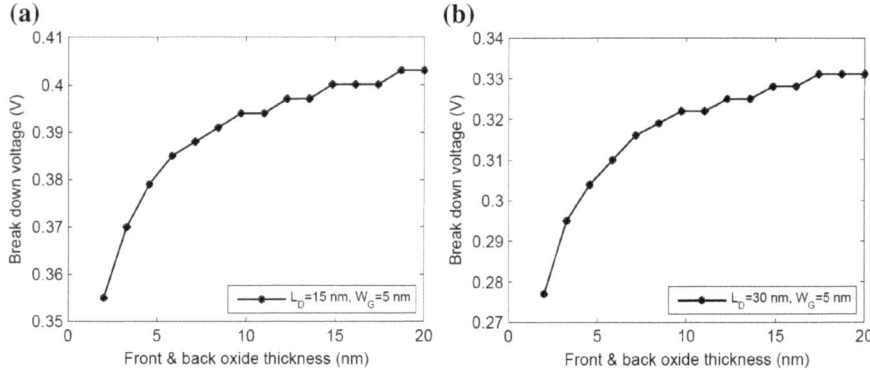

Fig. 4.20 Breakdown voltage for double-gate graphene nanoribbon transistor at different oxide thickness and channel length

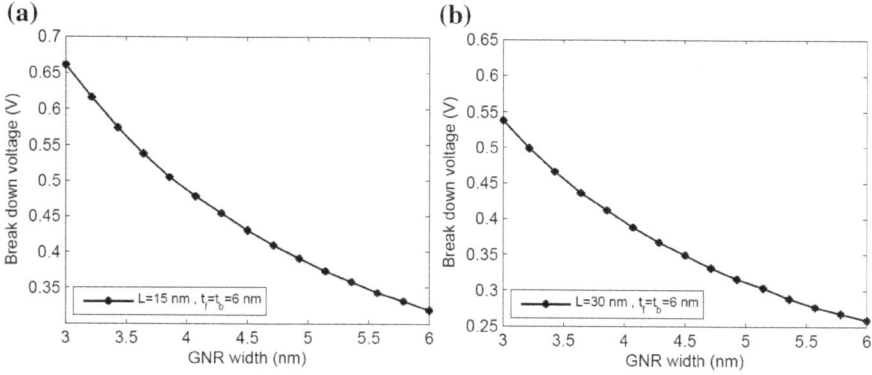

Fig. 4.21 Breakdown voltage for DG-GNRFET at different GNR's width and channel length

should be decreased. In addition, this figure shows that shorter channel length results in higher breakdown voltage. Therefore, in order to achieve high breakdown voltage, it is required to use short channel, narrow GNR and thick oxide. In addition, the dependence of breakdown voltage on the gate-source voltage of double-gate device is shown in Fig. 4.22. As a conclusion, a simple formulation as a guideline for optimization of breakdown voltage can be given as

$$BV \propto \frac{\alpha t_{ox} L_d}{W_g \cdot L \cdot V_g} \qquad (4.14)$$

As can be seen in Fig. 4.21a, b, the operating voltage of double-gate GNRFET could be limited to a level as low as 0.25 V which could be even lower than threshold voltage depending on the oxide thickness and the material used as oxide. In addition, this figure shows that the most effective parameter in controlling

Fig. 4.22 Breakdown voltage of single-gate GNRFET against gate voltage

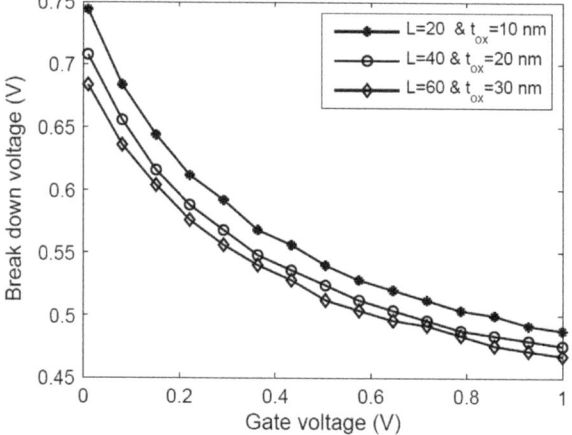

maximum operating voltage of GNRFETs is the channel width. Lowering channel width increases both the bandgap and breakdown voltage. However, it is worth mentioning that it also impairs the carrier mobility due to edge scattering. Therefore, there is a trade-off between breakdown voltage and device delay, which has to be taken into account for designing GNRFETs.

4.3.3 Comparison Between Breakdown Voltage of DG- and SG-GNRFETs

This section presents results of breakdown simulation of single- and double-gate GNRFETs together for conducting a comparison study. The common and interesting point concluded from studying Fig. 4.23a–c is that the breakdown voltage of a double-gate GNRFET is less than a counterpart device at the same condition. It was depicted before that length of velocity saturation region, where breakdown happens, in double-gate GNRFET is less than that of single-gate GNRFET. As it is known, breakdown voltage is calculated from $\int_0^{L_d} \alpha = 1$.

Therefore, it is clear that breakdown voltage decreases with L_d. In addition, in Fig. 4.23a, it is shown that both BV of DG-GNRFET and BV of SG-GNRFET are functions of channel width. However, the effect of channel width is more in single-gate device. The situation is different when the variable parameter is channel length. As can be seen in Fig. 4.23b, breakdown voltage of single gate device is almost independent of channel length when it is compared to that double-gate device. However, as the channel length gets longer than almost 20 nm, there is small change in breakdown voltage as channel length increases. Finally, Fig. 4.23c shows that the oxide thickness slightly influences the BV of single- and double-gate GNRFETs. In conclusion, the most effective parameters in adjusting breakdown voltage of single- and double-gate GNRFETs are GNR's width and channel length, respectively.

4.3.4 Comparison with Silicon Counterpart Devices

Our aim is not to compare the breakdown voltage of GNRFET with counterpart silicon-based FETs because there is no data for breakdown voltage of silicon-based FETs with one-atom-thick silicon channel. However, we show the existed experimental data for 40, 50 and 100 nm FET compared with a GNRFET at the same channel length, oxide thickness and gate-source voltage *but not the same channel thickness and width*. Table 4.2 shows the results.

It can be said that since ionization coefficient of GNR could be more than that of silicon (depending on GNR width), assuming same drift region for GNRFET and silicon-based FET, the breakdown voltage of silicon-based device is more than that

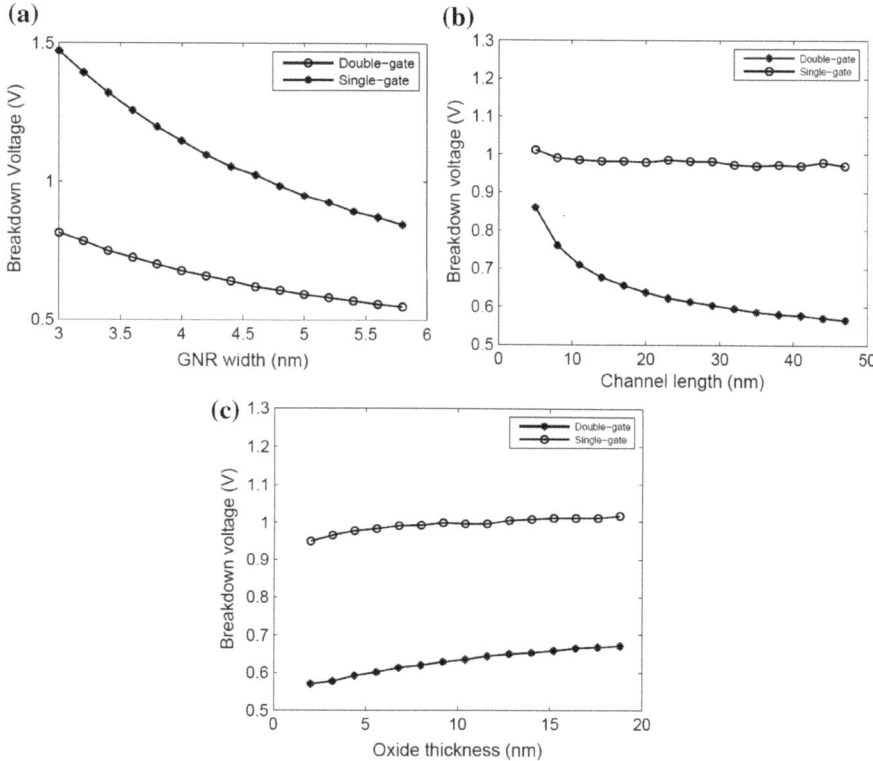

Fig. 4.23 A comparison between breakdown voltage of single- and double-gate GNRFETs at different GNR's width, channel length and oxide thickness. It shows that breakdown voltage of a double-gate device is almost half the BV of single-gate counterpart

Table 4.2 Breakdown voltage of GNRFET compared to existed experimental data for silicon FETs in different technologies

Device	Technology	Channel length (nm)	Breakdown voltage (V)	Reference
Single-gate	GNR	40	0.98	This work
Double-gate	GNR	40	0.62	This work
Single-gate	Si-SOI	40	2.11	[6]
Single-gate	GNR	50	0.57	This work
Double-gate	GNR	50	0.47	This work
Single-gate	Si-bulk	50	6	[7]
Single-gate	GNR	100	0.51	This work
Double-gate	GNR	100	0.42	This work

of a counterpart GNRFET. This later characteristic could be considered as another drawback for GNR in a case that high-voltage application is required. However, it can be exploited in avalanche photodiodes (APDs) which function based on impact ionization process and demand very high speed (100 Gbps) due to their use in telecommunication. In APDs, high voltage is applied to increase multiplication factor which is given as

$$M(x) = \frac{1}{1 - \int_0^{L_d} \alpha(F) \mathrm{d}x} \qquad (4.15)$$

where F is electric field. Clearly, using graphene in APDs with a higher α than that of silicon results in higher multiplication factor at the same operating voltage. Therefore, operating voltage of these sensors which are as high as 200 V can be suppressed. In addition, due to high mobility of graphene, device speed could be improved. The model presented in this thesis could be applied to study the behaviour and optimization of these kinds of devices.

4.4   Validation Range of the Proposed Model

This model has been proposed for GNR with width in range of up to 10 nm according to definition of GNR [8]. The ionization threshold voltage has been considered in range of $4E_g$ to $6E_g$ according to Gaussian simulation. In addition, the verification of results shows that the model agrees well with simulation for electric field in order of 10^6 V/m.

References

1. W. Yang, X. Cheng, Y. Yu, Z. Song, D. Shen, A novel analytical model for the breakdown voltage of thin-film SOI power MOSFETs. Solid-State Electron. **49**(1), 43–48 (2005)
2. K. Yeom, J. Hinckley, J. Singh, Calculation of electron and hole impact ionization coefficients in SiGe alloys. J. Appl. Phys. **80**(12), 6773–6782 (1996)
3. M. Ghadiry, M. Nadi, M. Saiedmanesh, H. Abadi, An analytical approach to study breakdown mechanism in graphene nanoribbon field effect transistors. J. Comput. Theor. Nanosci. **11**(2), 339–343 (2014)
4. M. Ghadiry, A.A. Manaf, M.T. Ahmadi, H. Sadeghi, M.N. Senejani, Design and analysis of a new carbon nanotube full adder cell. J. Nanomater. **2011**, 36 (2011)
5. M. Frisch, G. Trucks, H.B. Schlegel, G. Scuseria, M. Robb, J. Cheeseman, G. Scalmani, V. Barone, B. Mennucci, G. Petersson, Gaussian 09, Revision A. 02. Gaussian, *Inc., Wallingford, CT*, 200 (2009)
6. V. Su, I. Lin, J. Kuo, G. Lin, D. Chen, C. Yeh, C. Tsai, M. Ma, Breakdown behavior of 40-nm PD-SOI NMOS device considering STI-induced mechanical stress effect. IEEE Electron Device Lett. **29**(6), 612–614 (2008)

7. H. Wong, Drain breakdown in submicron MOSFETs: a review. Microelectron. Reliab. **40**(1), 3–15 (2000)
8. M.T. Ahmadi, Z. Johari, N.A. Amin, A.H. Fallahpour, R. Ismail, Graphene nanoribbon conductance model in parabolic band structure. J. Nanomater. **2010**, 12 (2010)

Chapter 5
Conclusion and Future Works on High-Voltage Application of Graphene

5.1 Summary and Conclusion on Breakdown, Ionization and Surface Potential of Graphene Field-Effect Transistors

Shrinking transistor sizes has been the most feasible and effective approach to reduce power and delay of MOSFETs for decades. However, by reaching the nanoscale dimensions, silicon is facing limitations for downscaling such as short-channel effects. As a result, new device concepts such as graphene FETs are being introduced as alternatives to silicon. Since graphene has a zero bandgap, graphene nanoribbon of this material has been introduced to open a bandgap, which was the focus of this study.

This thesis studied the ionization and breakdown mechanism of single- and double-gate graphene nanoribbon FETs (GNRFETs) using analytical approach and modelling. There were three main variables in this work, length of saturation velocity region, L_d, ionization coefficient, α, and breakdown voltage, BV, itself. The L_d is a solution of surface potential at saturation region; therefore, surface potential was modelled by applying Gauss law on GNR channel. Then, α was found to be function of lateral electric field and probability of collision through ballistic and drift motions. Therefore, ionization coefficient was analytically modelled using lucky drift approach and deriving equations for drift and ballistic motions. By developing the model of α, breakdown voltage was calculated using avalanche breakdown condition, $\int_0^{L_d} \alpha \, dx = 1$.

Using the proposed models, simulation was done by MATLAB, and the effect of several structural parameters such as oxide thickness, channel length and GNR's width was studied on the breakdown voltage, length of saturation velocity region and impact ionization coefficient. It was shown that operating voltage limit for GNRFETs could be as low as 0.25 V. However, after optimization, it could be increased to 1.5 and 0.8 V in single- and double-gate GNRFETs, respectively.

In addition, it was found that altering channel length does not significantly influence the breakdown voltage of single-gate device as it did in conventional

© The Author(s) 2018
I.S. Amiri and M. Ghadiry, *Analytical Modelling of Breakdown Effect in Graphene Nanoribbon Field Effect Transistor*, SpringerBriefs in Applied Sciences and Technology, https://doi.org/10.1007/978-981-10-6550-7_5

devices while it is influential in that of a double-gate device. Furthermore, the profile of breakdown voltage with respect to oxide thickness and channel width was plotted, and the effective parameters on BV were summarized as follows

$$\text{BV} \propto \frac{\alpha t_{\text{ox}} L_{\text{d}}}{W_{\text{g}} \cdot L \cdot V_{\text{g}}} \tag{5.1}$$

We showed that the most effective parameter in controlling maximum operating voltage of GNRFETs is the channel width. Lowering the channel width increases both the bandgap and the breakdown voltage. However, it is worth mentioning that it also impairs the carrier mobility due to edge scattering. Therefore, there is a trade-off between them. What is more, it was revealed that BV of double-gate GNRFET is less than that of single-gate GNRFET at the same condition meaning that double-gate GNRFET is not a suitable candidate for high-voltage application. In terms of ionization coefficient α, it was found that α is strongly dependant on GNR's width and is structurally tunable as shown in

$$\alpha \propto \frac{F(x)}{E_{\text{g}} \cdot W \cdot E_{\text{t}}} \tag{5.2}$$

As the result showed, α in GNR could be four times more than that of silicon at a same electric field. This property of GNR could be interpreted as desired or undesired characteristic based on the application.

In addition, a rough comparison between breakdown voltage of conventional devices and GNRFETs showed that GNRFET's breakdown is less than that of the silicon-based devices. We have modelled both L_{d} and α in this project; therefore, the approach presented here could be used in study, design and optimization of specific devices which function based on impact ionization process such as avalanche photodiodes (APDs) or high-power devices based on GNR. In addition, they could be further extended to study the threshold voltage, short-channel effects and current in GNRFET or even bilayer graphene FET with slight modifications.

5.2 Future Works on Breakdown Mechanisms and High-Voltage Applications of Graphene

In terms of fabrication, it is interesting to fabricate a prototype device suitable for measuring substrate current and calculation of ionization coefficient to verify the presented theoretical results in this work. Furthermore, considering the effect of parasitic capacitances in surface potential model could be a further work, and finally, application of the approach together with conductance model in sensors to study the GNR-based sensors such as GNR-based DNA sensor or APDs is a fruitful research to do.